I0799407

IMAGES
of America

ESCAMBIA COUNTY

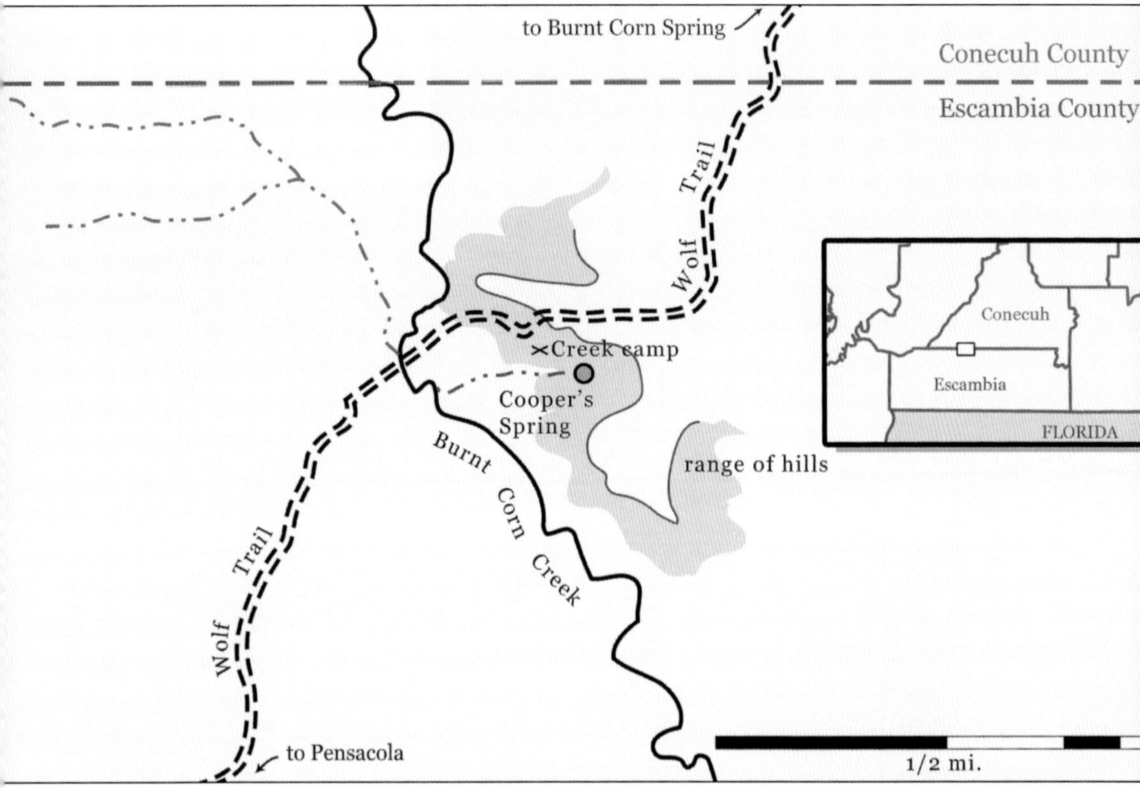

The single most important event to take place in Escambia that had national significance was the Battle of Burnt Corn Creek. This battle took place on July 27, 1813, when American troops, consisting of whites, mixed-blood Indians, and full-blood Indians, attacked a group of Creek Indians known as Red Sticks on the Wolf Trail. The Creeks were taking arms back to their villages in the Upper Creek area of Alabama to be used against American settlers and troops. The Creeks expelled the American forces that day. The Battle of Burnt Corn Creek was the beginning of the Creek Indian War that ended on March 27, 1814, at the Battle of Horseshoe Bend. This battle brought national attention to Gen. Andrew Jackson. (Courtesy of Mitchell McMillan and Malcolm McMillan.)

On the Cover: Aubrey Strong, Dan Brantley (holding the fish), and Joe Larkin (right) proudly display their catch in the Pollard area on the Conecuh River. Hunting and fishing have always been a source of enjoyment in Escambia County. In the early days, residents could use a net and bring home a bucket full of fish. The county has many sloughs, lakes, and swamps associated with the Conecuh River bottom. (Courtesy of D. W. McMillan Trust.)

IMAGES
of America

ESCAMBIA COUNTY

Tom McMillan on behalf of the
Escambia County Historical Society

ARCADIA
PUBLISHING

ISBN 978-0-7385-9123-0

Published by Arcadia Publishing
Charleston, South Carolina

Printed in the United States of America

Library of Congress Control Number: 2011936444

For all general information, please contact Arcadia Publishing:
Telephone 843-853-2070
Fax 843-853-0044
E-mail sales@arcadiapublishing.com
For customer service and orders:
Toll-Free 1-888-313-2665

Visit us on the Internet at www.arcadiapublishing.com

This book is dedicated to US Army Sgt. Tamara Thurman, of Brewton, who lost her life on that tragic day in 2001, when the 9/11 terrorist flew into the Pentagon.

Contents

ACKNOWLEDGMENTS

When I agreed to put together this book, I decided that it would be a good time to collect as much local material as I could to save for future generations, much like so many have done in this area in the past. I want to thank all of those who contributed photographs, articles, time, and ideas to this effort. I hope to pull all of this together and donate it to the Escambia County Historical Society for safekeeping. Many images that were contributed did not make it into the book.

I want to thank those people who went beyond the call. That group would include Jerry Fischer, Yank Lovelace, Robert Thrower, Jerry Simmons, David Allen, Kevin McKinnley, Carolyn Conn, Phillip Parker, Jim Hildreth, and Thomas Lawson Jr. I relied on the help of others to do some of the legwork, scanning, and organizing. That group would include Megan Brantley, Mandy Holland, Mitchell McMillan, Malcolm McMillan, and Roy Brown. It was especially fun to work with my sons Mitchell and Malcolm. Randy Tatano helped me wrap up some loose ends in Atmore, and Canoe and was a joy to work with.

I cannot thank all of those who contributed, so please forgive me. I wish I could have used hundreds of other pictures, but in the end I had to use those that fit the theme and those that I had enough information about to write a good caption.

Introduction

Escambia County was created by an act of the Alabama Legislature in 1868. Most of the area was part of Conecuh County, but when the courthouse was moved from Sparta to Evergreen, the people in the southern part of the county along the state line decided that they wanted their own county. Part of Baldwin County was taken in along with the southern part of Conecuh County to form Escambia County. Before the formation of Escambia, this area had been part of Georgia, then the Mississippi Territory, and finally the Alabama Territory. Before all of the boundaries and borders were drawn, the Escambia County area was part of Creek Indian lands. There were no major Native American villages, but the Creeks used the land for hunting and gathering, as evidenced by signs of small, seasonal villages that have been found in the area.

Once the Indians began to trade with the new settlers, they also began to intermarry. In the 1790s, a group of 16 Creek Indian families who were of mixed blood and starting a new life moved from the upper Creek villages of the Alabama River region to Tensas in Baldwin County. In the 1820s, they moved to the Poarch community north of Atmore, stayed together as a band, and were officially known as the Creek Nation east of the Mississippi. Some of these families were the McGillivarys, Weatherfords, Tates, and Monicas. On August 11, 1984, this community was recognized by the federal government and is the only Indian tribe recognized in Alabama.

As the French developed their settlement around Mobile, Alabama, the Spanish around Pensacola, Florida, and the English in Georgia, trade began to develop with the Native Americans. Agents would live with the Indians, and more American families began to move into the area. As the roads and rivers carried more settlers, however, more conflicts began to develop. In 1811, the Federal Road from Georgia to Mobile opened, and land became available from the Federal Land Office. Eventually, hostilities with the British came to a head with the War of 1812. As a result, troop movement and port activity picked up in the Gulf Coast region. The British were using their agents with the Indians to supply them with arms to make trouble with the settlers and with any military post as part of their strategy to win the war. News of these activities resulted in the Battle of Burnt Corn, whereby the local militia of about 180 men organized in July 1813 to intercept a large contingent of Creek Indians that were carrying arms and ammunition from Pensacola to central Alabama by way of the Wolf Trail Trading Path. The Spanish and the British were helping to arm the Native Americans. The militia, headed by Col. James Caller, initially drove the Indians out with a surprise attack and took possession of the arms. After things settled down, however, the Creeks rallied and drove off the militia. The mission failed, and the troops dispersed. This was the beginning of the Creek Indian War, where the Wolf Trail crossed Burnt Corn Creek in Escambia County, and is the single most important event in the county's history.

On August 30, 1813, Creek warriors known as Red Sticks attacked Fort Mims in Baldwin County and killed several hundred mixed-blood Creeks, settlers, and militiamen. Another result of the Spanish helping the Native Americans was the establishment of Fort Crawford in Escambia County in 1816. Maj. Gen. Andrew Jackson, commander of the US Army of the South, did not appreciate what the Spanish were doing, so he established several military posts along the Florida line to protect the settlers and ultimately take control of Spanish Florida. Fort Crawford, located on a bluff above Murder Creek in what is now East Brewton, was one of Jackson's forts. Ultimately, General Jackson used the difficulty of supplying the fort and several other atrocities on the Federal Road as an excuse to take Pensacola and Florida away from the Spanish. Once General Jackson had run the Spanish military out of Pensacola and back to Cuba, the State Department took over and eventually paid Spain for Florida in 1821. During this time, Fort Crawford diminished in purpose until it was eventually abandoned.

As the area opened up to more settlers, the next big development was the railroad. The ports of Mobile and Pensacola needed to be connected to the rest of the country, and the rails came right through Escambia County. In 1861, the Alabama & Florida (AF) Railway was completed, connecting Pensacola and Montgomery. The community known as William's Station, which would later be known as Atmore, came into being on that railroad. Soon thereafter, in 1862, the Mobile & Great Northern Railroad was completed, and it connected with the AF in Pollard. After the Civil War, the connection was relocated to Flomaton, making that area of the county the rail center and giving it a big economic boost. During this period, the Escambia County area had still been part of Conecuh County, but when the courthouse was moved from Sparta to Evergreen, the residents near the state line petitioned the state legislature to create a separate county. In 1868, Escambia County was created out of Conecuh and Baldwin Counties. Pollard was chosen as the county seat. Around this time, demand for timber had increased because of a peaceful Europe as well as the Industrial Revolution. Cities began to grow in Europe, and housing for the workers caused a big need for timber. As it turned out, the seemingly limitless virgin forest near the Gulf Coast was ideal to supply the demand—the timber boom was on.

In the 1870s and 1880s, large mills were built in the Atmore and Flomaton areas. W.M. Carney and his son-in-law H.H. Patterson built the Carney Mill Company in Atmore. Sullivan brothers Martin and Dan, who had served as Union soldiers in the area, returned to Flomaton to make their living in timber. They teamed up with Pensacola banker Martin Sullivan and began to buy timberland in Alabama, ultimately acquiring 300,000 acres. Eventually, they would build and run Alger-Sullivan Lumber Company, one of the largest sawmills in the South.

Escambia County was not part of the plantation economy of Alabama, but it did have a strong early role in the timber economy. Numerous mills were built, and fortunes were made. Since bringing the logs to the mills required a lot of labor and organization, logging camps were established all over the county. Oxen, rivers, log ditches, and eventually logging railroads transported the logs to the mills. With jobs came all the other needs: supplies, shops, schools, churches, and stores. As the timber was cut, some of the cutover land was utilized for crops, such as strawberries, tobacco, and cotton. Many mill owners cut their timber and moved on, while others stayed and replanted pine trees.

As Alabama and the Gulf Coast grew, so did Escambia County. Its natural resources played a big part in that growth. Besides timber, Escambia County's other natural resources spurred economic growth. The demand for turpentine, for example, created more jobs and opportunities in the county. Local clay was proven to be of suitable quality for making bricks. The Keego Brick Company has had a long history of making and supplying brick. The sand and gravel deposits in the Flomaton area became the basis of another industry. Finally, oil was discovered at Pollard in 1952. Escambia County is located on huge deposits of oil and gas, which are still being drilled and produced today all across the county.

Today, Escambia County is home to over 38,000 people. It has a diverse economy and many recreational activities, such as hunting, fishing, golf, and the beautiful beaches along the Gulf. We are proud of our communities and look forward to the future.

One

Those Who Came First

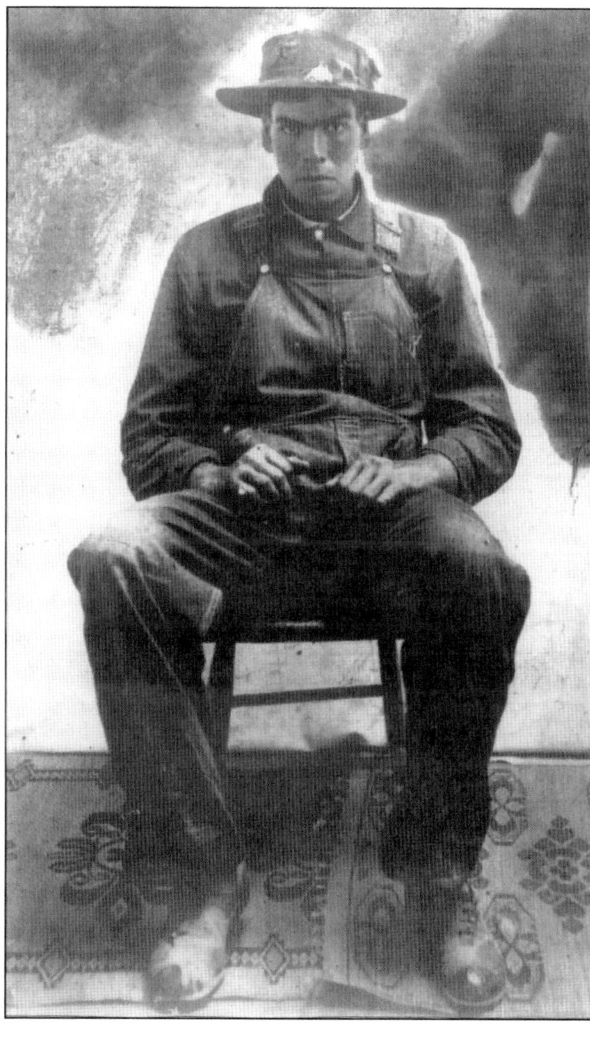

Enon Gibson was a member of the Hog Fork Indian community and was born in 1893. This photograph was taken about 1910. The local band of Creeks had moved to Tensaw in Baldwin County in the 1790s from the Montgomery area and later settled in the Atmore area in the 1820s. Gibson was a member of one of the original 16 families that made the move. (Courtesy of Marie Gibson/Dr. Tony Paredes.)

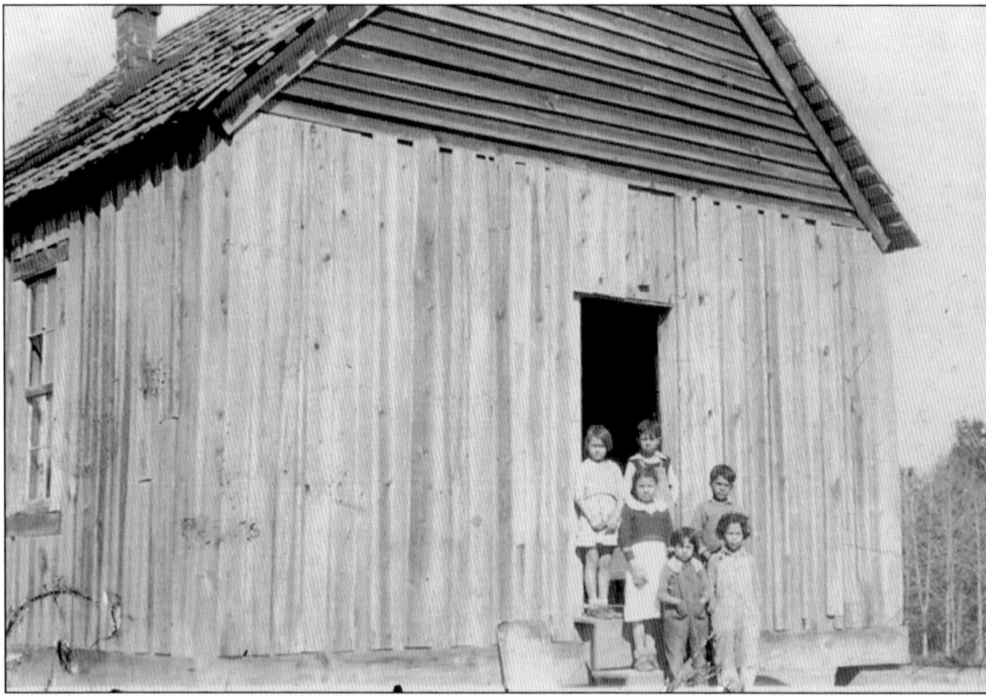

Students stand at the door of the Poarch Indian School. Several communities were settled by the Creeks. The Poarch community name became the surviving name and is now used to describe the whole Creek band in modern times. (Courtesy of Robert Thrower, Poarch Creek tribal historic preservation officer.)

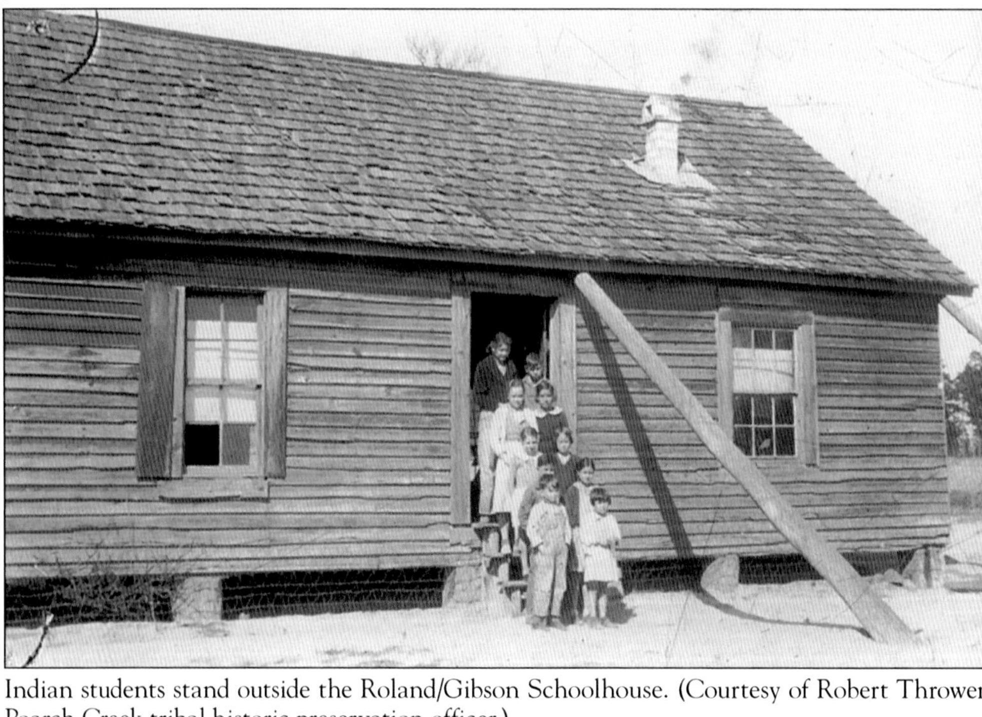

Indian students stand outside the Roland/Gibson Schoolhouse. (Courtesy of Robert Thrower, Poarch Creek tribal historic preservation officer.)

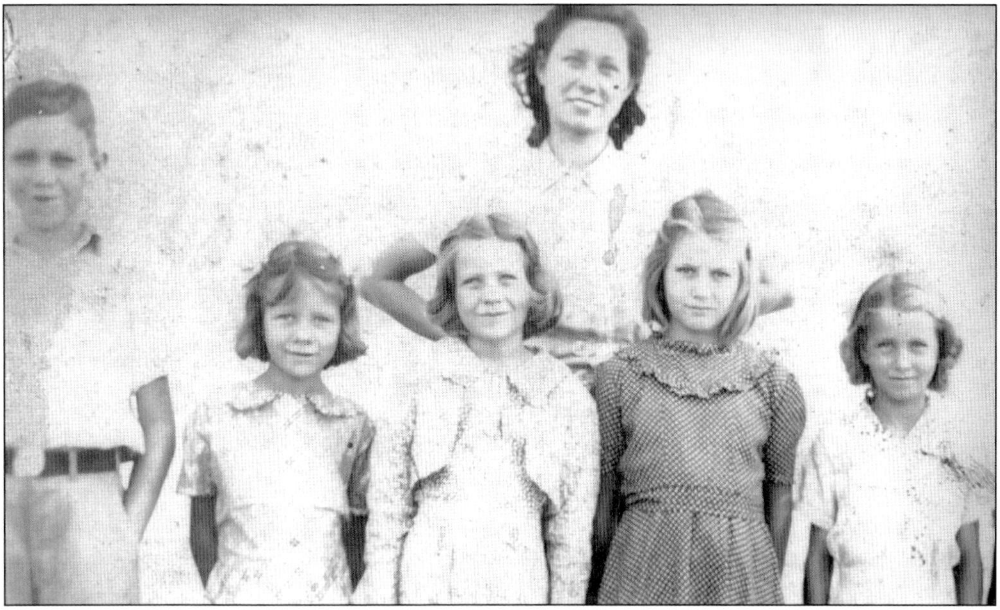

This photograph is of "Indian Children" from St. Mary's Episcopal Missionary Indian School, which was built in 1901. They are identified as all sharing one father and bearing the "Woods" surname. Today, little is known about St. Mary's School or the exact location of Pinedale, the recorded location of the school. It was the first church mission established through the Indian Mission Program. (Courtesy of Robert Thrower, Poarch Creek tribal historic preservation officer.)

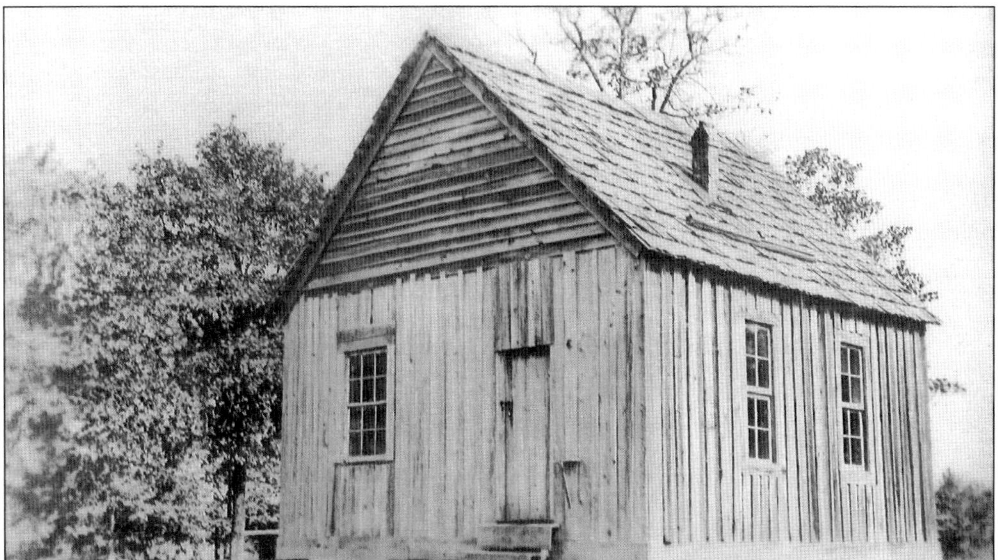

The old schoolhouse at the Headapadea community (also known as Perdido Hills) served as the location for services for the first Episcopal Church congregation within the community from 1929 to 1932. The name Headapadea stands for "Head of Perdido," as the headwaters of the scenic Perdido River were located within this tribal community. St. Anna's Mission Church was completed in 1932. (Courtesy of Robert Thrower, Poarch Creek tribal historic preservation officer.)

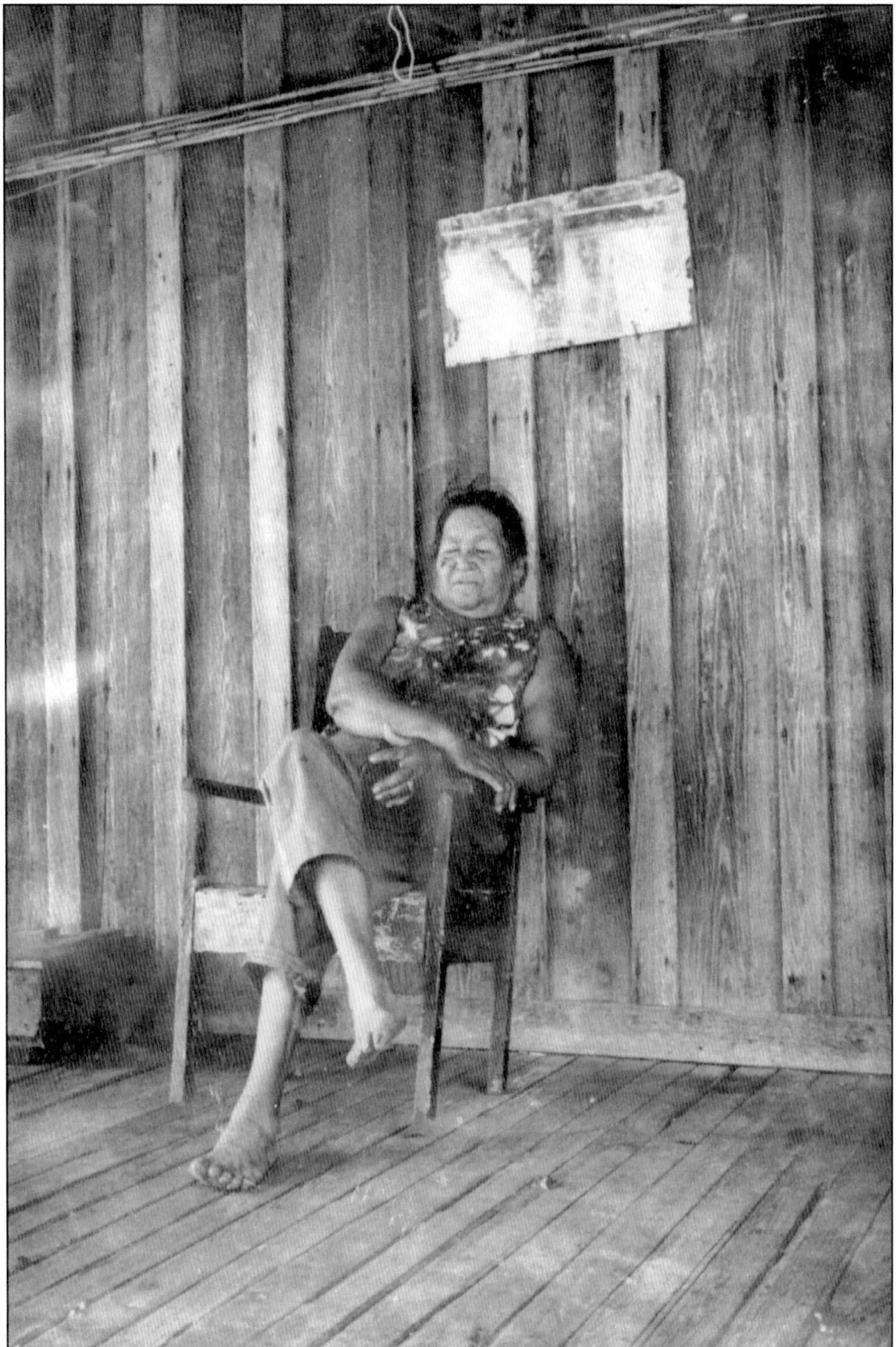

Lula Rolin relaxes on the front porch of her house in the Headapadea community. Of note is the mirror on the front porch, which some families put there to "keep the devil out of the house." It was believed that the nocturnal-roaming, highly prideful, and beautiful devil would be attracted to the mirror and then look at himself all night. He would admire his beauty until daylight, when he would then be forced to leave. (Courtesy of Robert Thrower, Poarch Creek tribal historic preservation officer.)

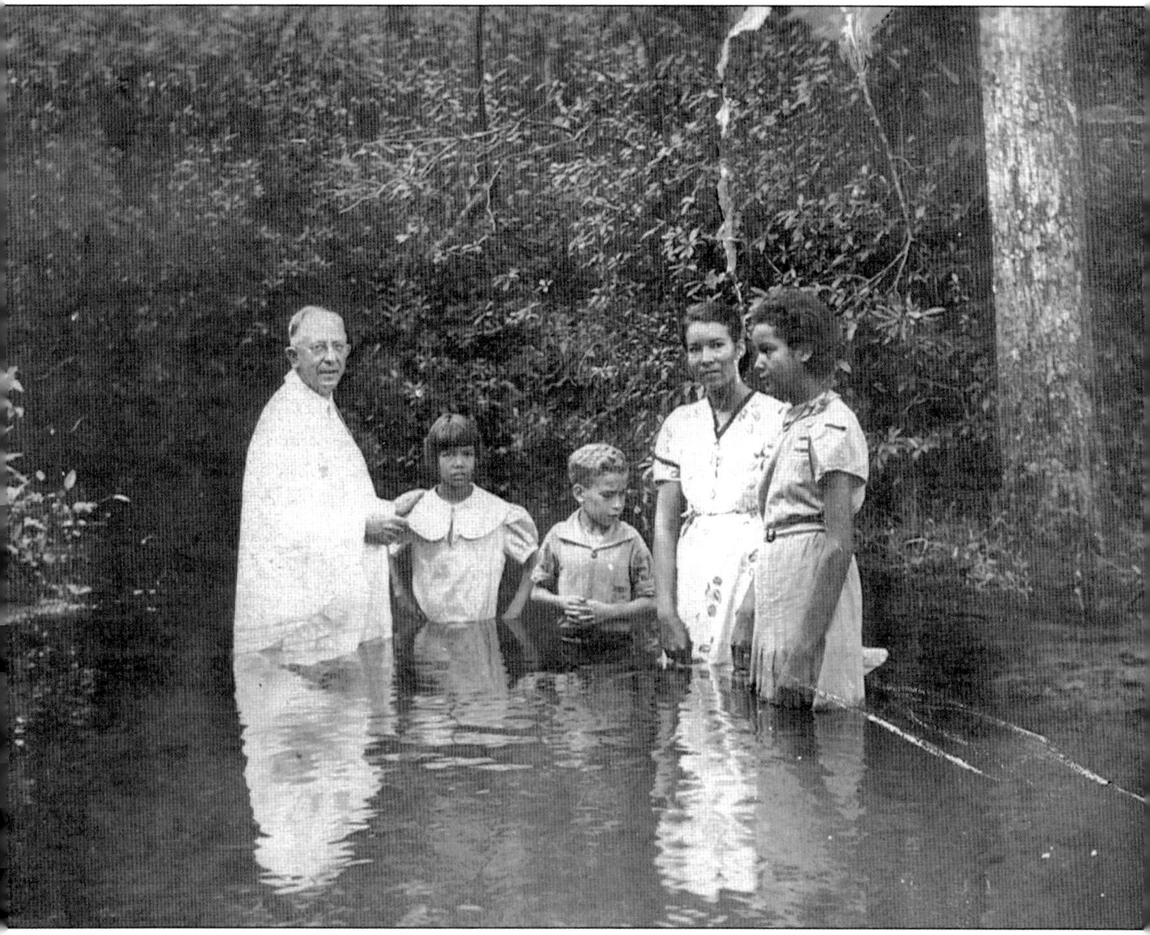

Rev. E. Van Edwards prepares to baptize members of his congregation in the local swimming hole at the Headapadea community. Reverend Edwards served the Poarch community from 1929 to 1943. In an excerpt from his notebook, Edwards writes, "We have about seventy members now and fully instructed in the Church's teaching and ways, no more loyal band of people anywhere to be found. Though very poor, they gave more in the Lenten offering last year, than three of my biggest white Mission." (Courtesy of Robert Thrower, Poarch Creek tribal historic preservation officer.)

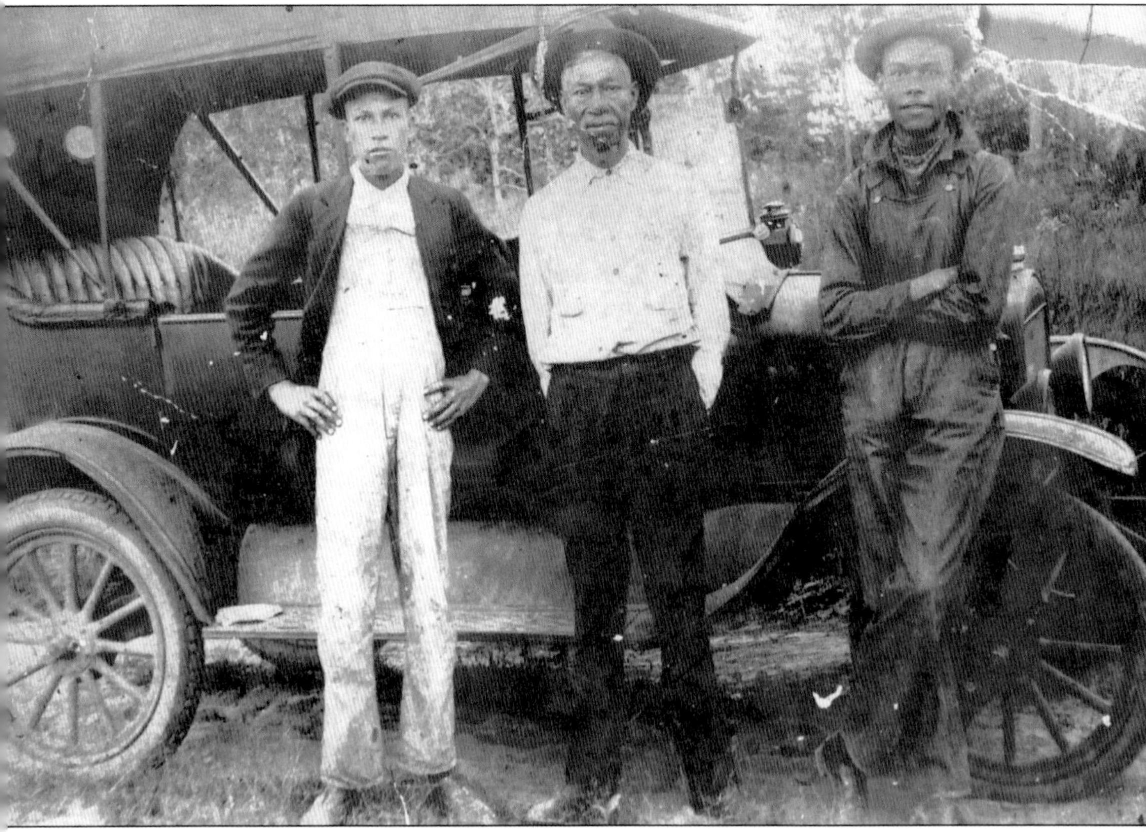

From left to right, Elbert, Bill, and Tracy Rolin stand in front of Anna Macy's Ford Model-T car. Following the death of medical missionary Dr. Robert Macy, Tracy drove his wife, Mrs. Anna, to various conferences in what would later be known as the "Missionary Model-T." St. Anna's Mission Church in the Headapadea community would soon be named after her. After Dr. Macy died, Anna asked Hattie Mae Hixon to live with her. Anna's home, being too remote for her to occupy, moved her residence to Atmore. A fine, trustworthy young man, Tracey became her chauffeur, a position that he is still very proud of. He says, "I was living in Poarch Switch, and I would catch the train there at six in the morning and ride into Atmore and drive Mrs. Macy around in her 'Missionary Ford' all day, wherever she wanted to go. I drove her even to two Diocesan Conventions . . . one in Florence, way up in the northern part of Alabama, and one in Birmingham. We sure did enjoy it. I was confirmed by Bishop McDowell at St-John's-in-the-Wilderness in 1935." (Courtesy of Robert Thrower, Poarch Creek tribal historic preservation officer.)

Chief Fred Walker, a notable community leader, passed away in 1941. He could not read, but he was very intelligent. Not only was he chief, but he was responsible for holding his people together, always working for their spiritual good. Episcopal Bishop McDowell licensed him to hold prayer meetings, but since he could not read he was not a lay reader. (Courtesy of Robert Thrower, Poarch Creek tribal historic preservation officer.)

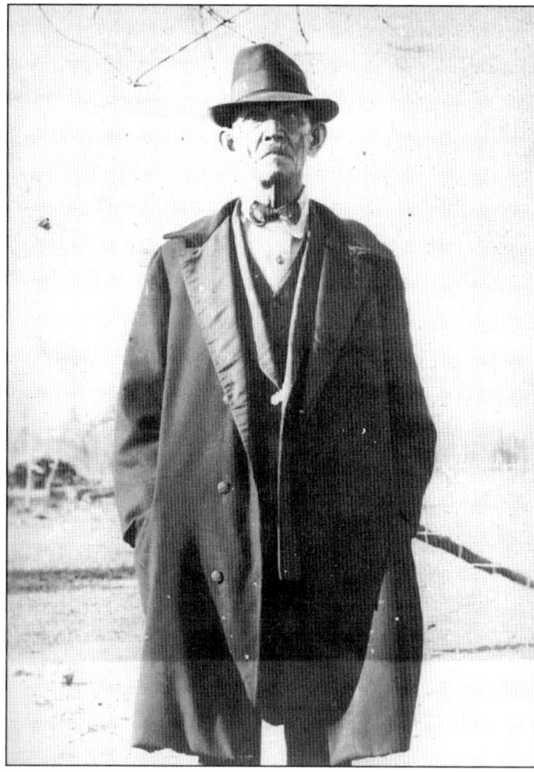

Jessie Freeman (left) and Jim Joiner display the basic tools of the timber trade: cross saw, axe, and the "bottle." The bottle contains kerosene and has been plugged with pine needles. While cutting pine trees, kerosene would sometimes be brushed on the cross saw blade to dissolve the pine resin building up on it and thereby provide lubrication to the cross saw. Jim Joiner was the grandfather of the late Nina Gail Thrower. (Courtesy of Joe Petty.)

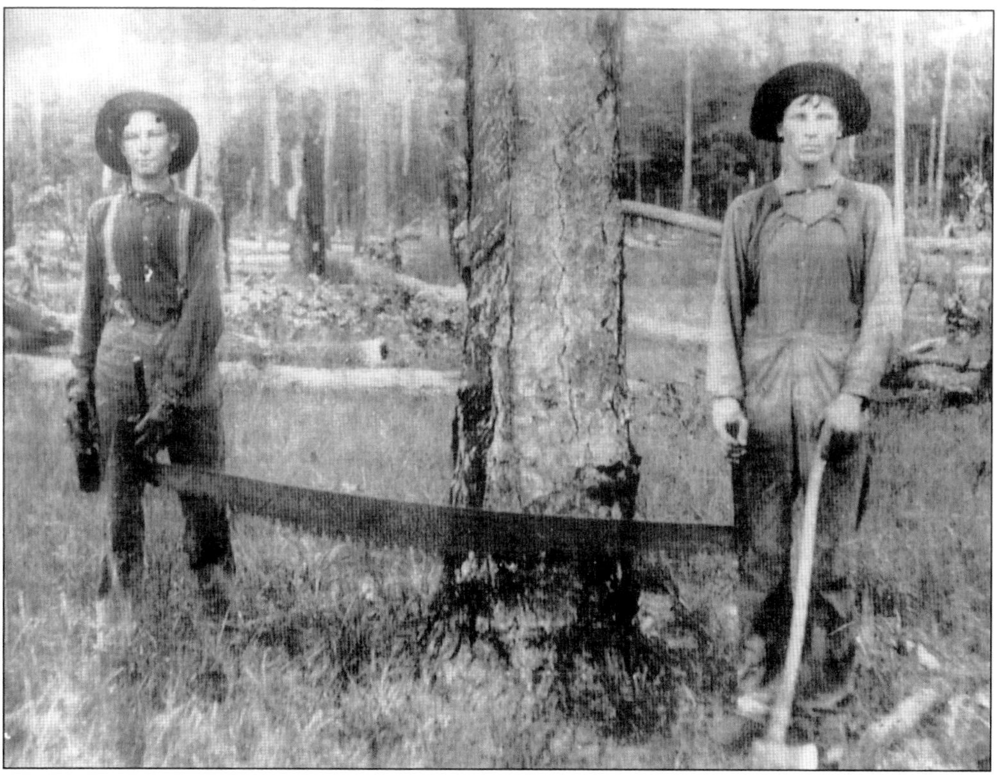

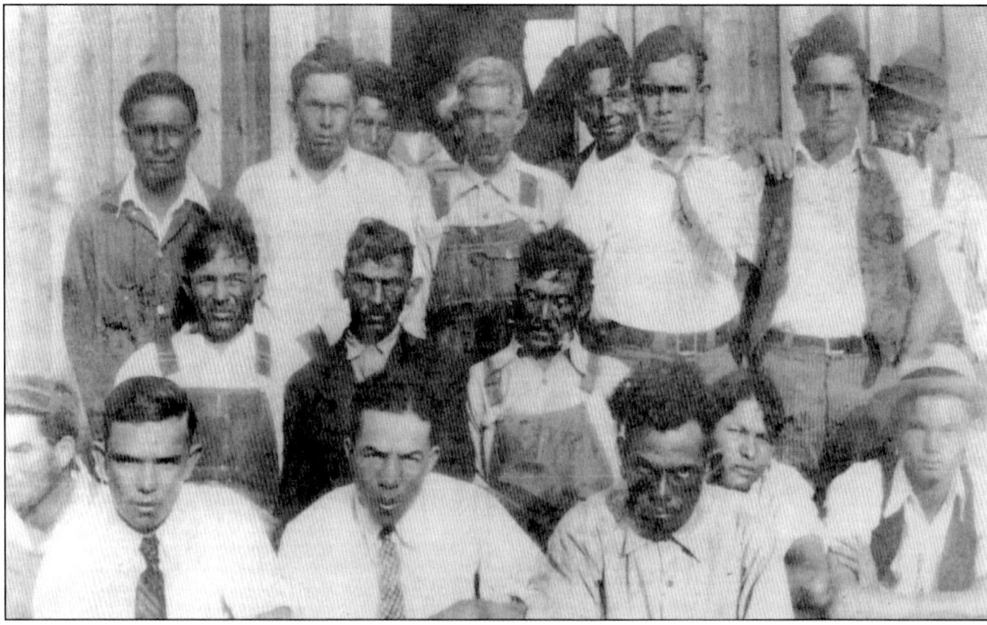

Some of the men from the tribal community built and completed St. Anna's Mission Church in 1932. Chief Fred Walker is second from the left in the second row. The church was named after Anna Macy while she was still living in honor of her dedication and service to Poarch Creek people. It is still active today and comprised of a largely Poarch Creek congregation. (Courtesy of Robert Thrower, Poarch Creek tribal historic preservation officer.)

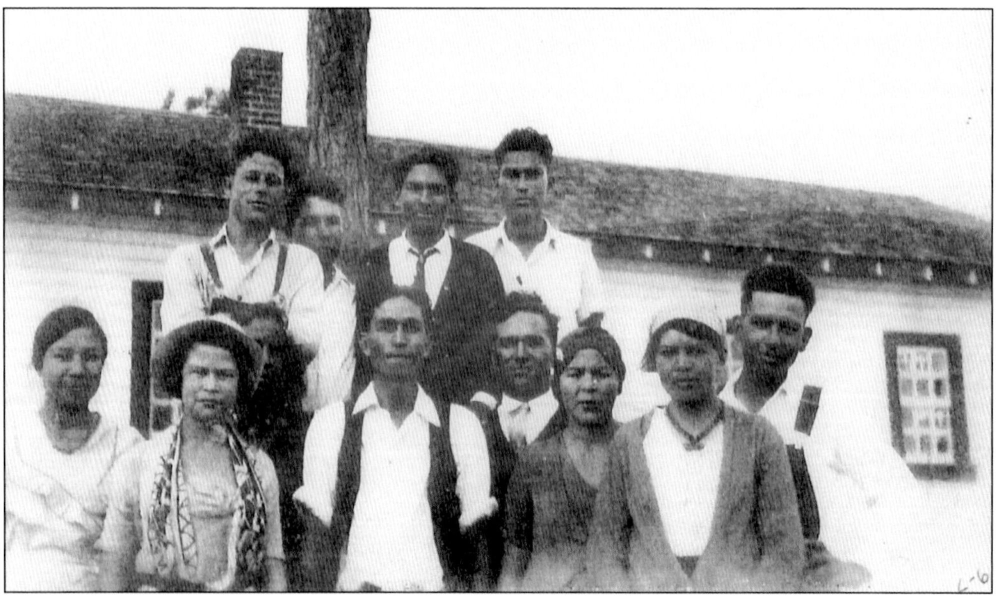

This photograph is of the Indian choir (about 1932) at St. John's in the Wilderness, an Episcopal mission church constructed in the Poarch Switch community. It was built concurrently with St. Anna's Church in 1932. These choir members are, from left to right, (first row) Loree Rolin, Elbie Presley, Margie Rolin McGhee, choir director Riley McGhee, Noah McGhee, Ella Lou McGhee, Lily McGhee, and Judge Rolin; (second row) organist J.D. Presley, Tommy McGhee, Mace McGhee, and Dunney Rolin. (Courtesy of Robert Thrower, Poarch Creek tribal historic preservation officer.)

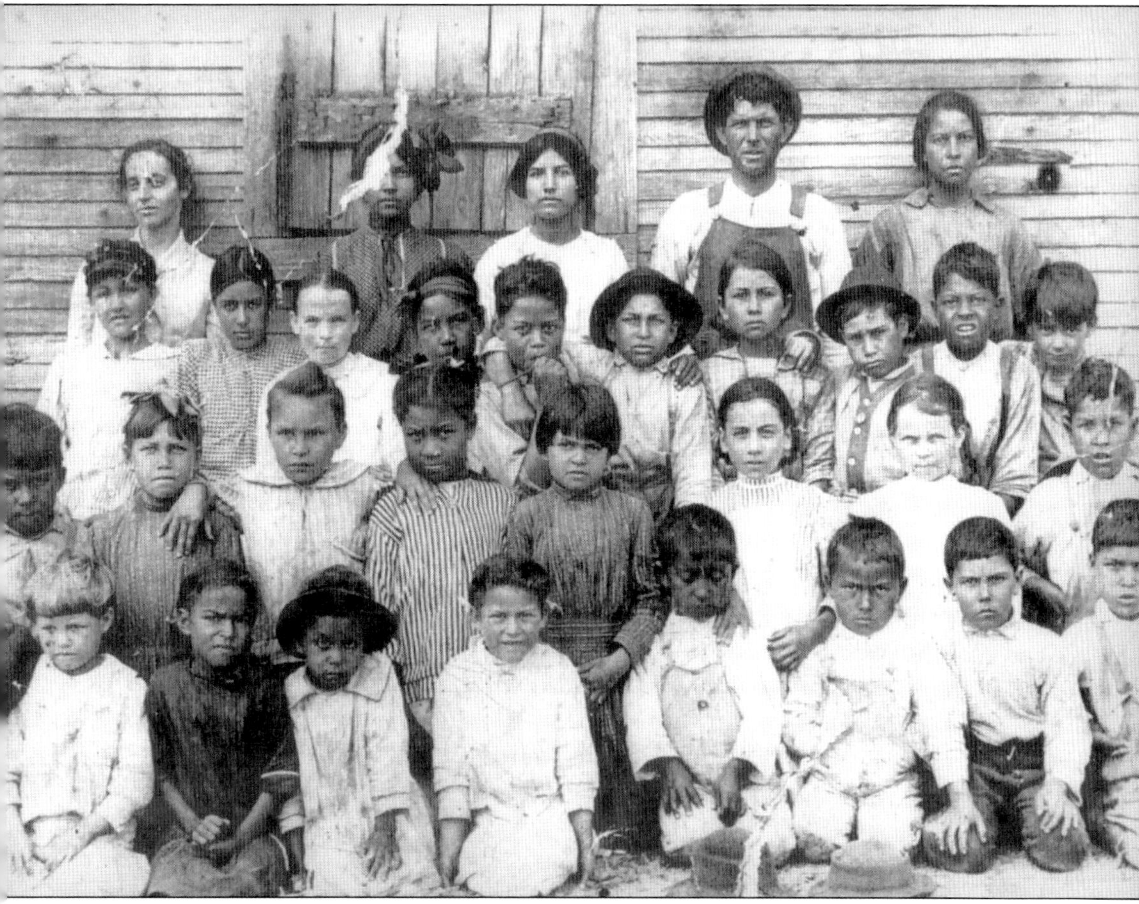

School students gather outside of the Hog Fork Community Indian School around 1910. A young Calvin McGhee, who would later become chief, is located in the center, wearing a hat with his left arm draped over the young Tracy Rolin (father of present-day chairman Buford Rolin). (Courtesy of Robert Thrower, Poarch Creek tribal historic preservation officer.)

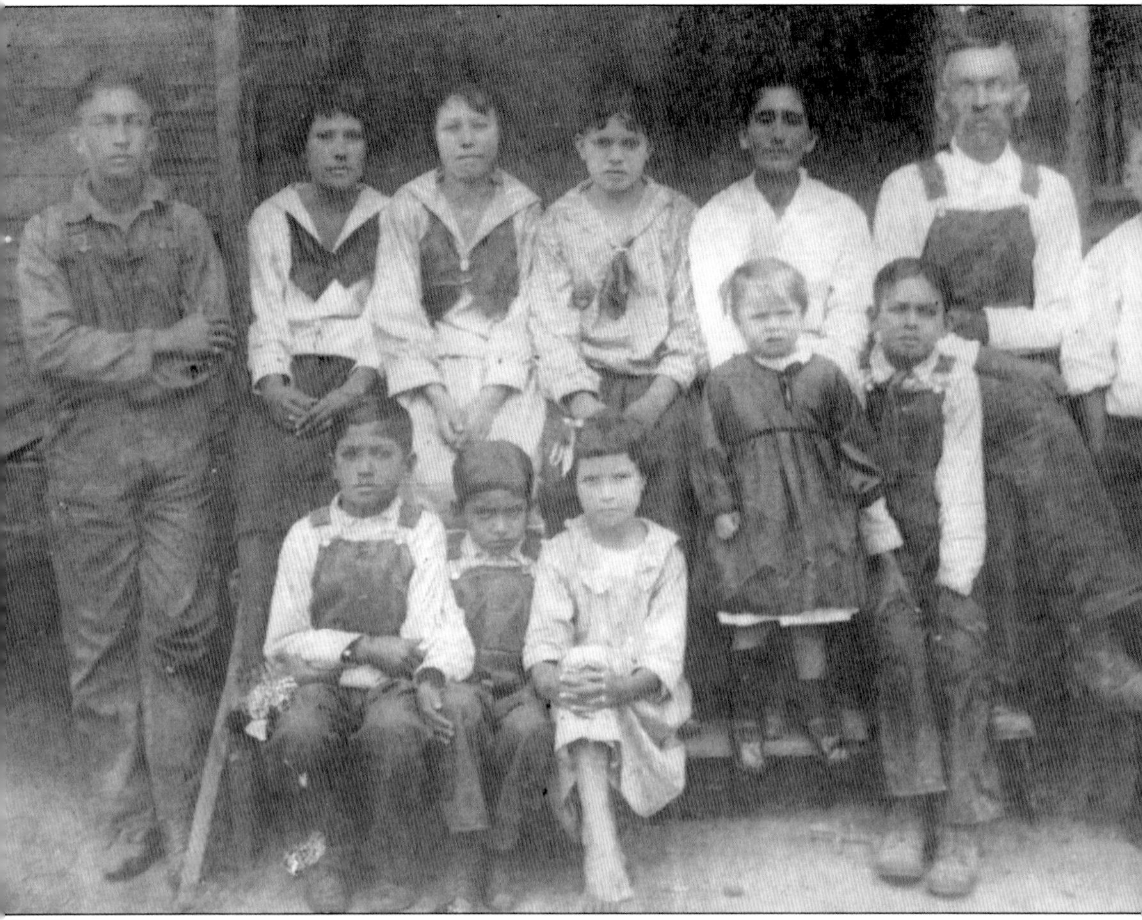

Chief Calvin McGhee (far left, second row) appears as a young man in this group family photograph. Calvin's mother, Ida McGhee, is located third from the right in the second row. Calvin's father, Lee McGhee, dressed in overalls, sits immediately to the right of Ida. (Courtesy of Robert Thrower, Poarch Creek tribal historic preservation officer.)

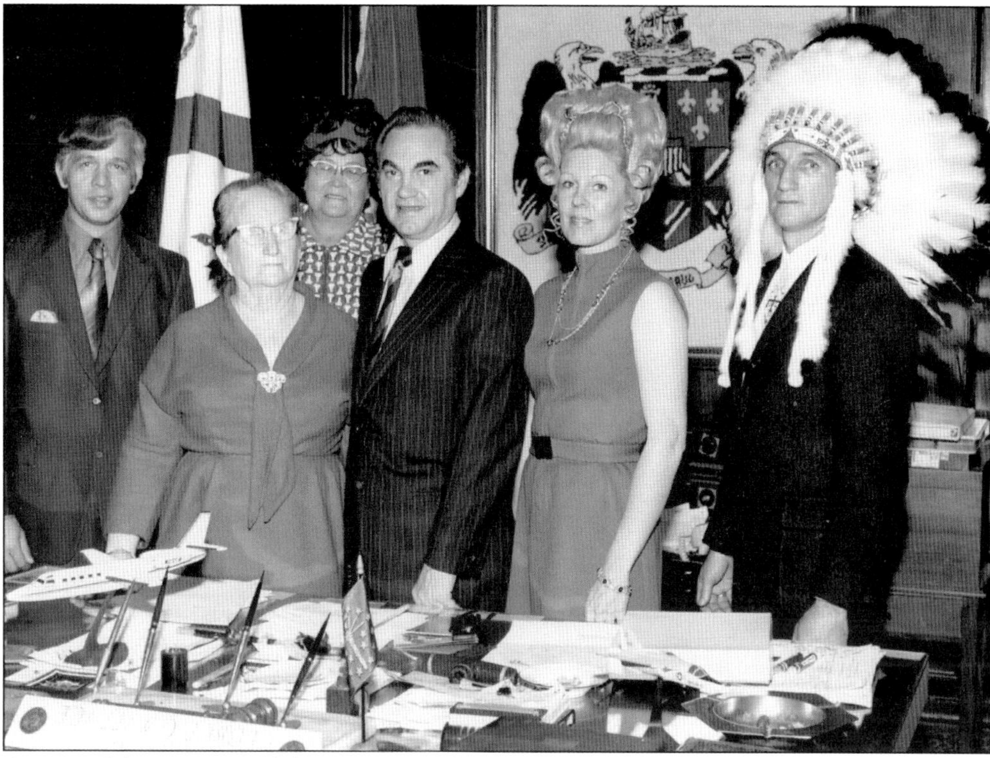

An east of the Mississippi delegation of Creek Nation visits Alabama governor George C. Wallace (center). Buford Rolin (current Poarch Creek tribal chairman) is on the left, and former chief Houston McGhee (son of Chief Calvin McGhee) is on the far right. (Courtesy of Robert Thrower, Poarch Creek tribal historic preservation officer.)

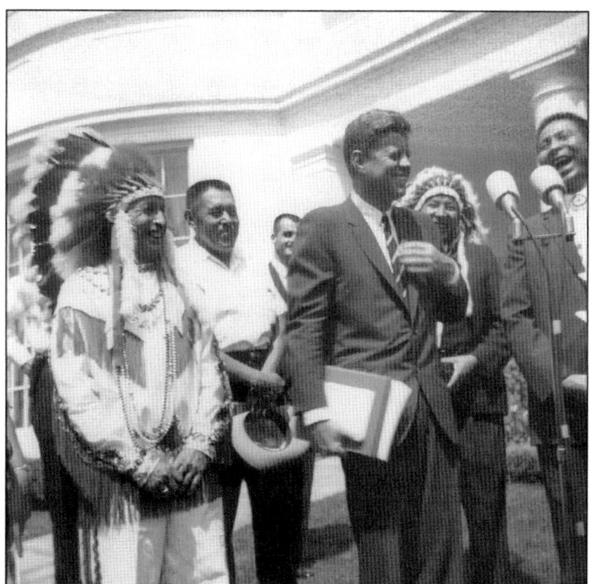

Taken in August 1962 outside the White House, this photograph shows Chief Calvin McGhee of the Poarch Creek Indians on the left, and Pres. John F. Kennedy appears in the center just before addressing the microphone. The others are not identified but are from various Indian tribes. They are all laughing at a reply Chief McGhee just gave to President Kennedy. Kennedy said that he had never seen a blue-eyed Indian before, and McGhee replied, "Well, you have now." McGhee died in 1970, some 14 years before the Poarch band of Creek Indians received federal recognition. (Courtesy of Robert Thrower, Poarch Creek tribal historic preservation officer.)

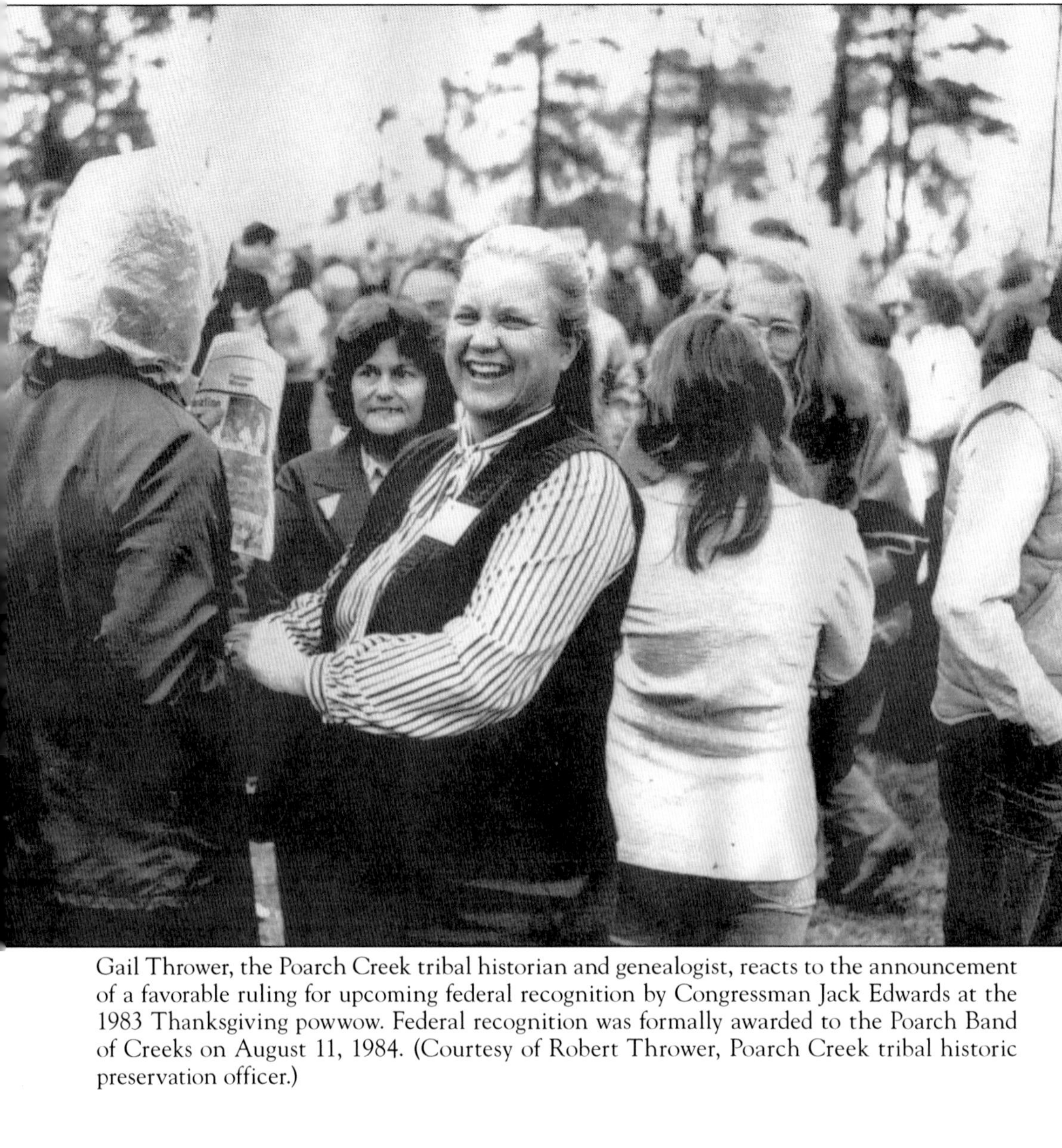

Gail Thrower, the Poarch Creek tribal historian and genealogist, reacts to the announcement of a favorable ruling for upcoming federal recognition by Congressman Jack Edwards at the 1983 Thanksgiving powwow. Federal recognition was formally awarded to the Poarch Band of Creeks on August 11, 1984. (Courtesy of Robert Thrower, Poarch Creek tribal historic preservation officer.)

Two

Timber Boom and the Iron Horse

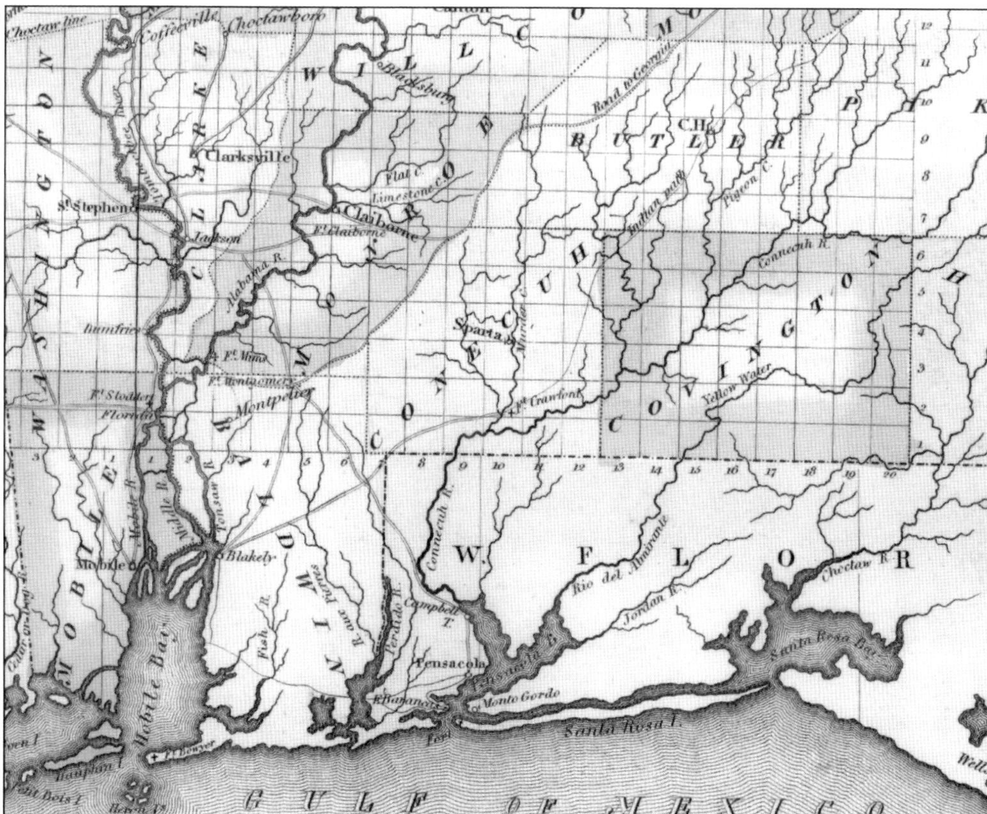

Alabama became the 22nd state in the Union in 1819. This 1823 map shows how the counties were formed in south Alabama. Escambia County was created by an act of legislature in 1868 by taking a large part of Conecuh County and a small part of Baldwin County. (Author's collection.)

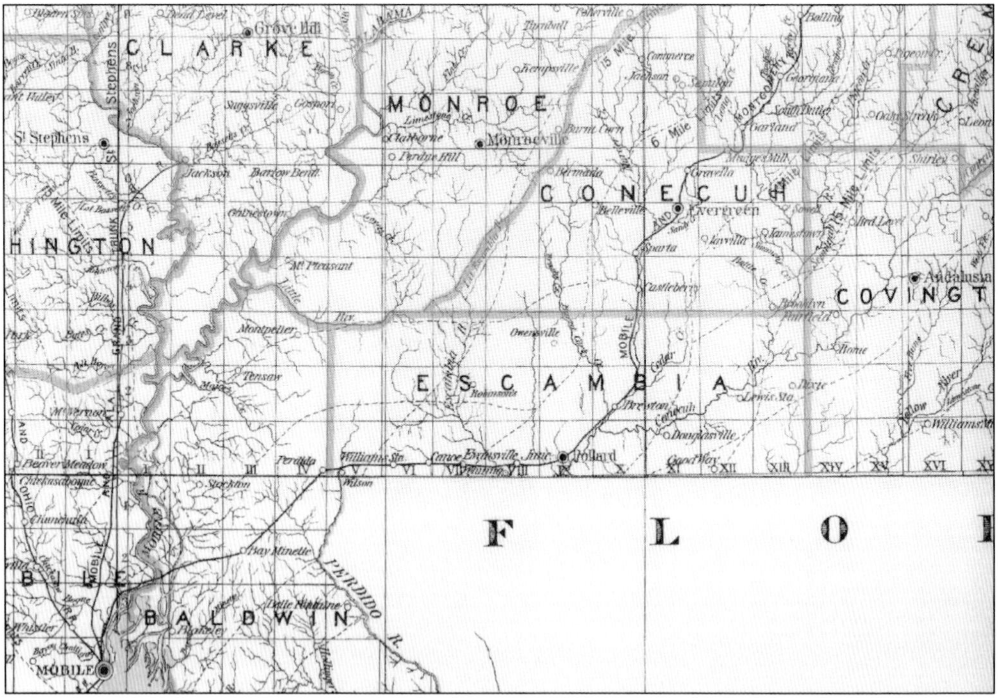

Because the county was created in 1868, this 1878 map accurately shows the principal communities at that time. Atmore was known as William Station, Flomaton was called Junction, and Pollard was the county seat. The railroad is shown except for the line that came from Pensacola into Junction. (Author's collection.)

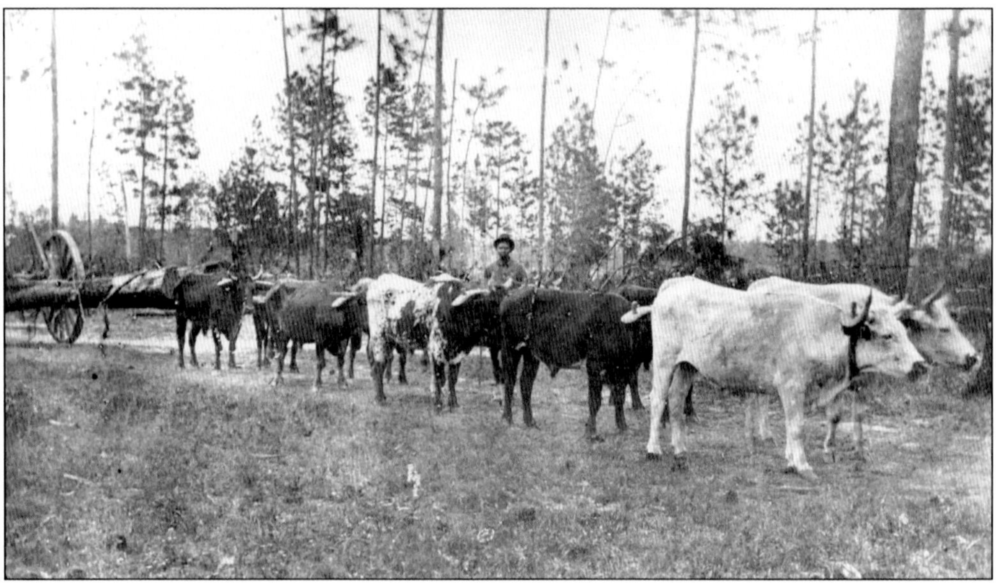

The earliest method of logging the virgin forest was by use of oxen teams. This setup of five oxen teams was typical; the steers nearest the log are called tongue steers, the second set are grab steers, the third set are swing steers, the fourth set are lead swing steers, and the final set are lead steers. The lead steers needed large white spots so the driver and other steers could see them in low-light conditions. (Courtesy of D.W. McMillan Trust.)

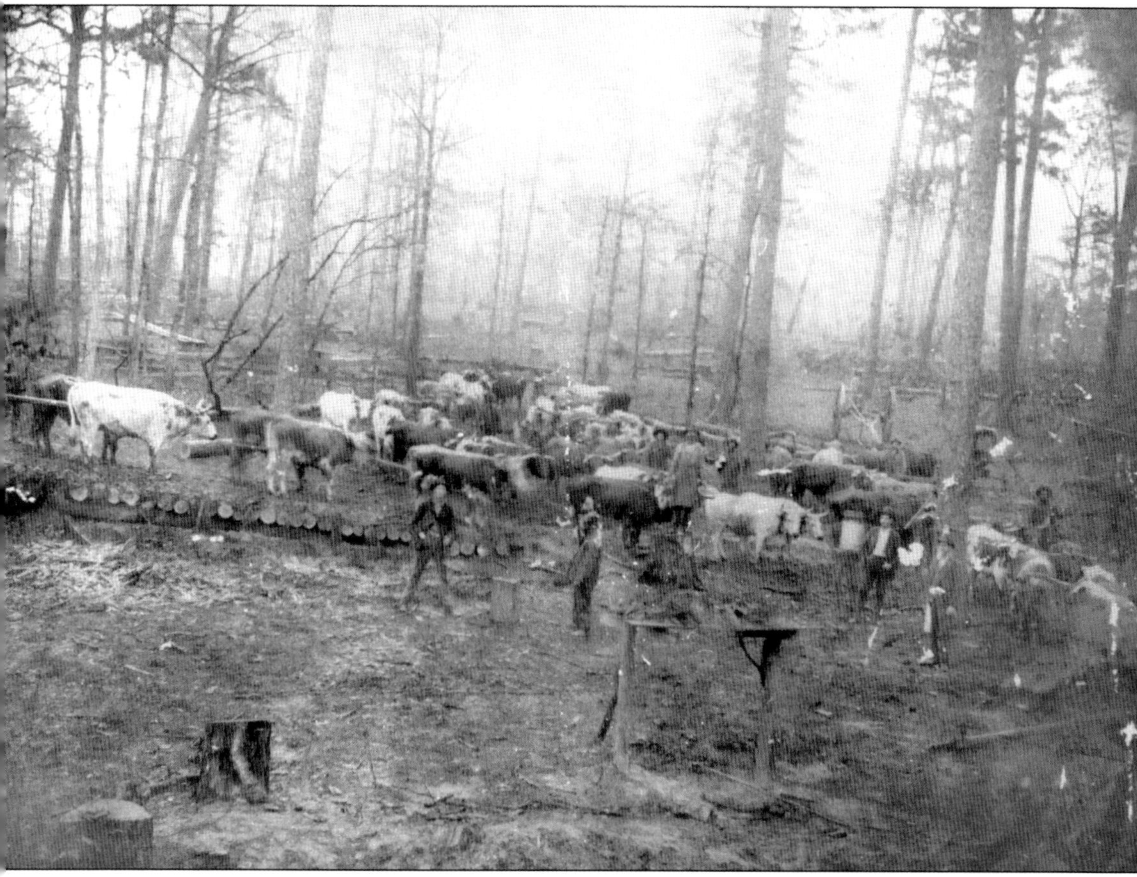

T.J. "Little Tom" Miller was the camp foreman of the Blacksher-Miller Lumber Company log camp near Belleville, Alabama. The picture was taken about 1894 around the same time that Miller was foreman. Before the timber was cut, it was virgin longleaf yellow pine. The timber was loaded on railroad log cars and carried to the London log pond, where it was moved by log ditch to the mill at Brewton. The log ditch was approximately six to eight miles in length. (Courtesy of D.W. McMillan Trust.)

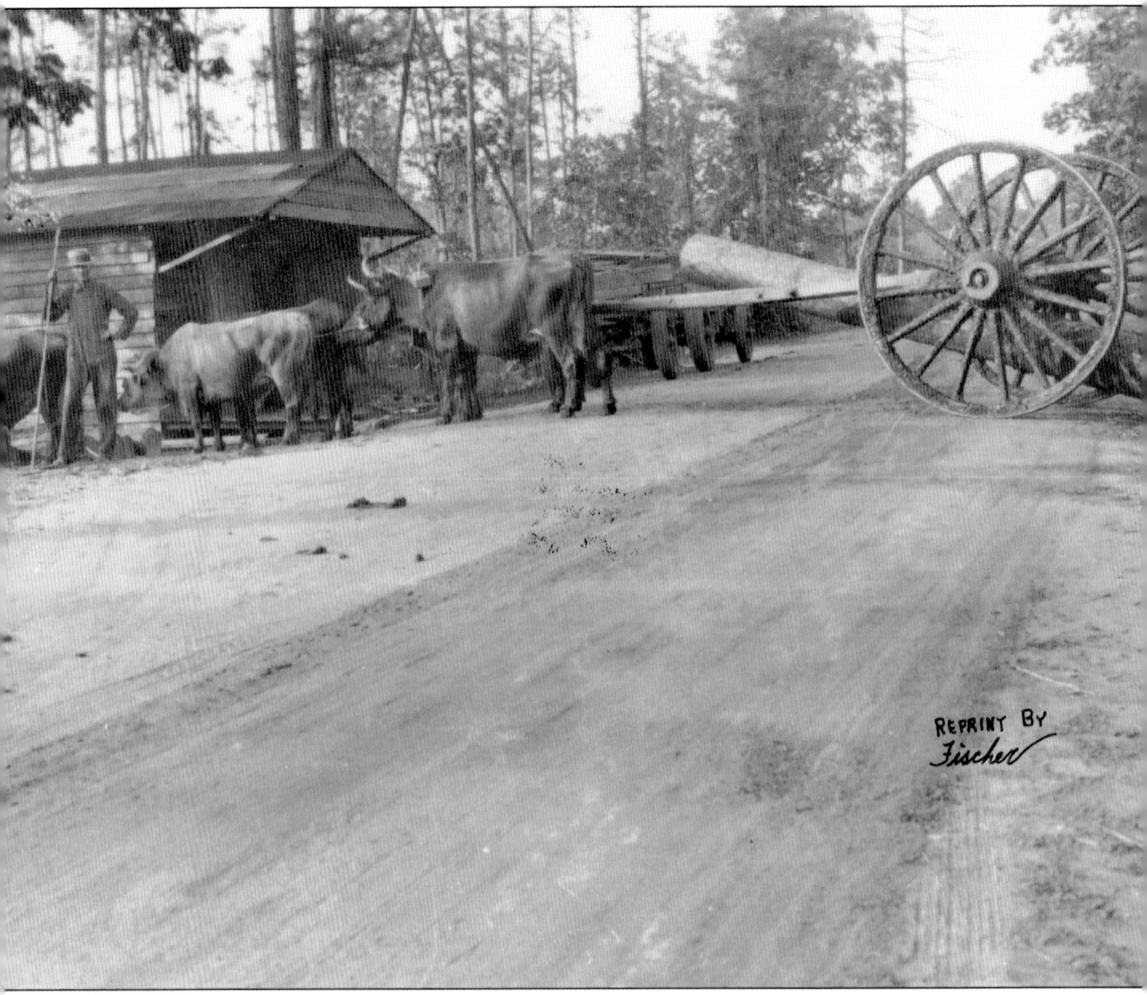

An oxen team skids a log into a staging area using a two-wheel cart called a caralog. The oversized wheels could cross wet and rough areas more easily. This scene was shot at Keego, where a railroad loading area had been established. (Courtesy of Jerry Fischer.)

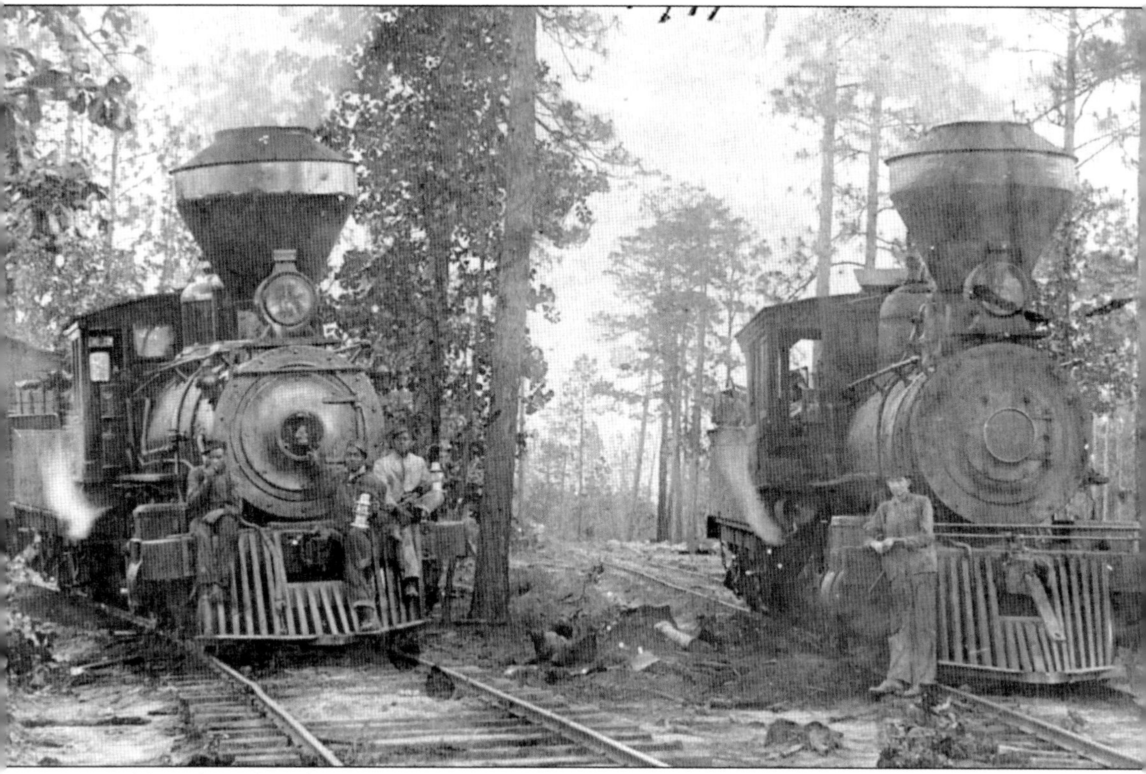

Two Baldwin funnel stock engines meet at Camp 8 just outside of McCullough (north of Atmore) in 1911. Alger-Sullivan Lumber Company of Century, Florida, owned large tracts of land in Escambia and Conecuh Counties. It established a railroad called Escambia Railway, which consisted of 90 miles of track, not including many miles of spurs. (Courtesy of Thomas Lawson Jr.)

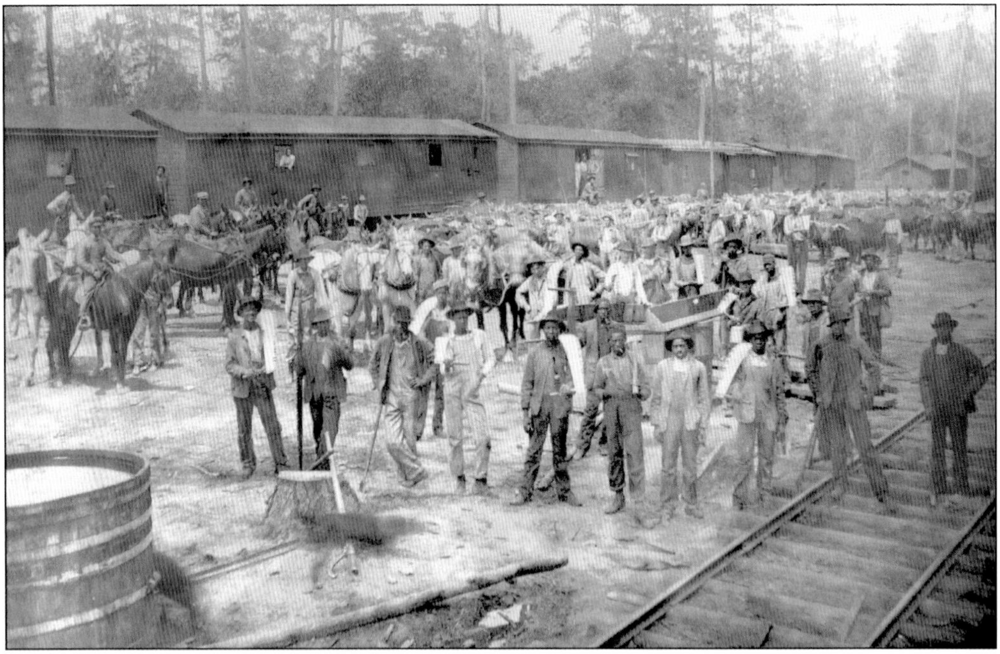

Camp 8 was owned and operated by the Alger-Sullivan Lumber Company. Such camps were known by their number, but they moved from place to place; therefore, it is difficult to know its location. Several of the men have crosscut saws resting on their shoulders. The logging railroad has pine logs for cross ties. This picture helps to demonstrate that a lot of jobs were created in the early days by logging the piney woods. (Courtesy of D.W. McMillan Trust.)

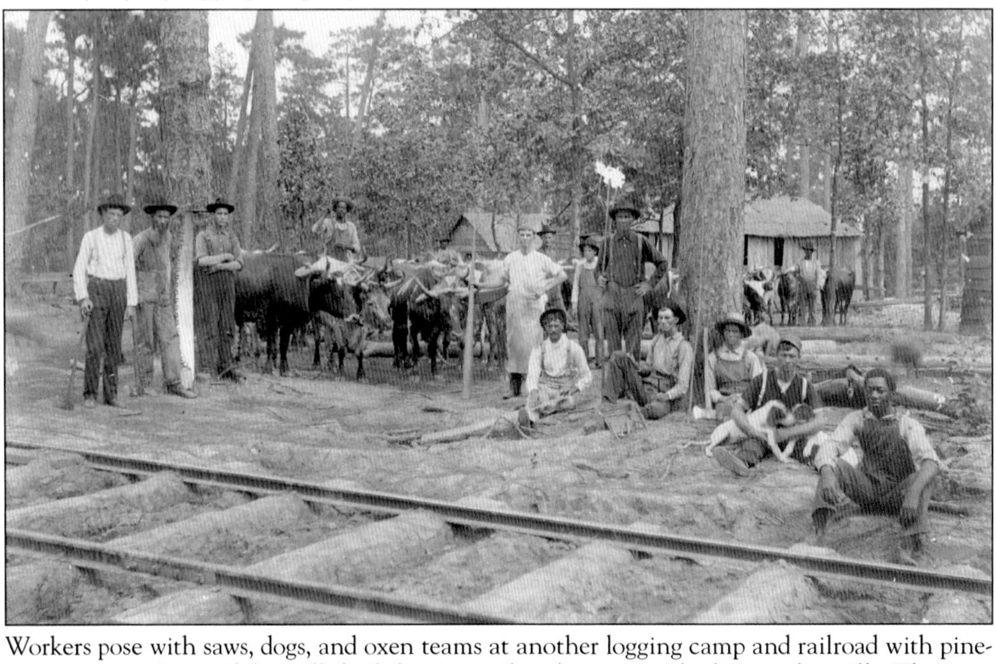

Workers pose with saws, dogs, and oxen teams at another logging camp and railroad with pine-tree cross ties. Most of the mills built logging railroads to move the logs to the mills. The trees in this photograph are large enough to be part of the virgin forest that was cut in the 1880s and 1890s for export to Europe. (Courtesy of D.W. McMillan Trust.)

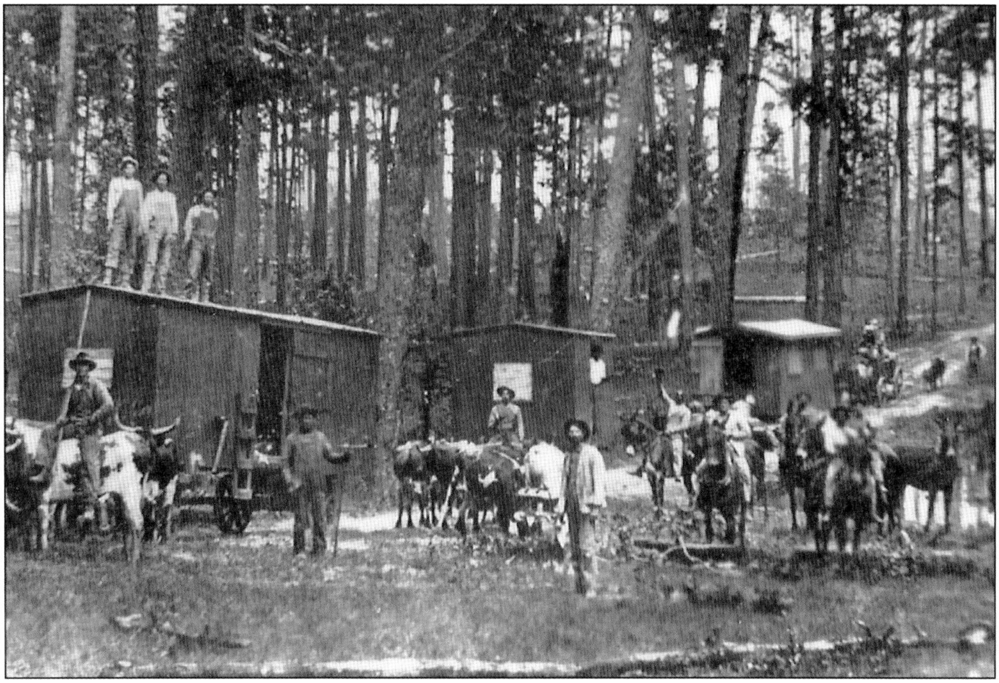

Logging camps were set up all over Escambia County. These horses, mules, and oxen are being used to move a camp from one location to another near Canoe. (Courtesy of Escambia County Historical Society.)

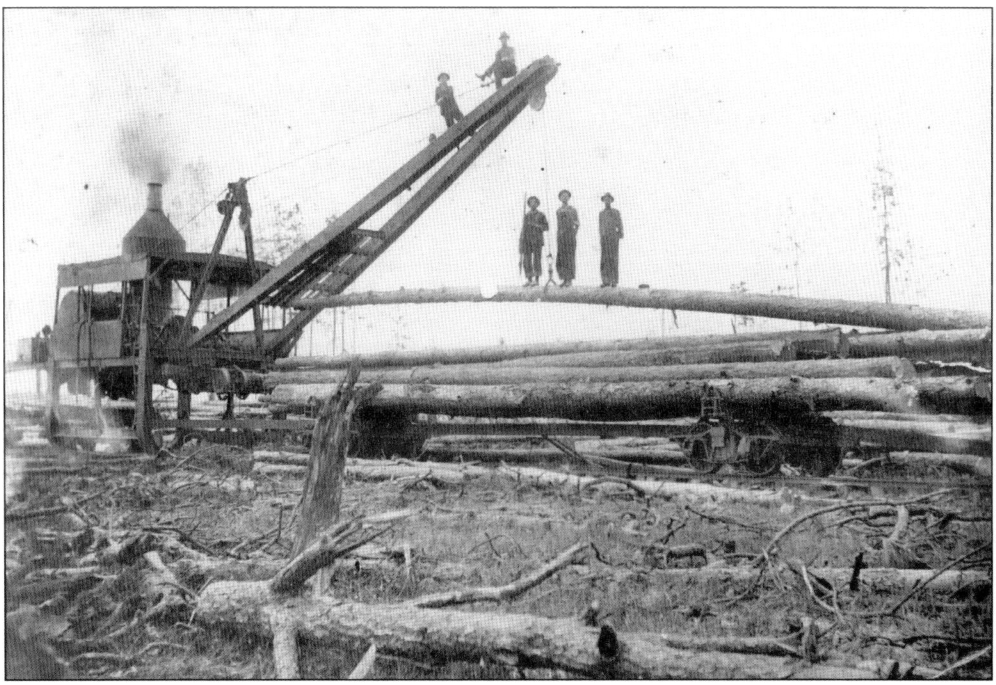

This early-model log loader, shown in the 1900s, lifted the logs by a winch that was driven by a steam piston. The boiler was typically heated by fat pine or lighter wood, which made a hot fire because it had a high resin content. (Courtesy of D.W. McMillan Trust.)

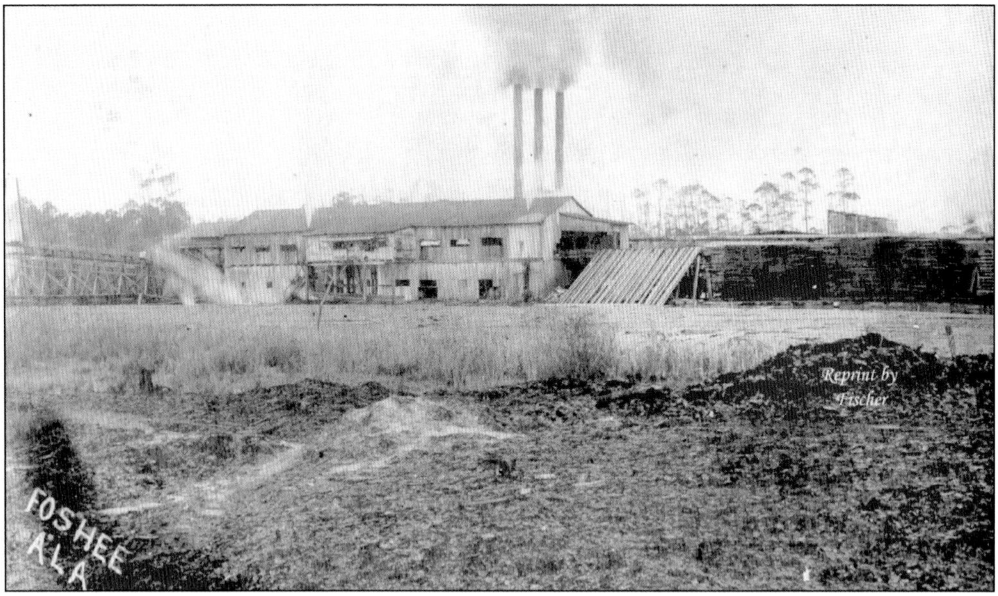

Located at Foshee, Alabama, Alger-Sullivan Lumber Company's first mill was called the Sullivan Mill before Gen. Daniel Alger and Pensacola banker Martin Sullivan joined forces to form Alger-Sullivan Lumber Company. The mill was later moved just across the Alabama-Florida state line and was enlarged to become one of the largest in the South around 1900. The new town that developed around that mill was named Century, because it was started at the beginning of a new century. This mill was finally closed for good in the early 1920s. (Courtesy of Jerry Fischer.)

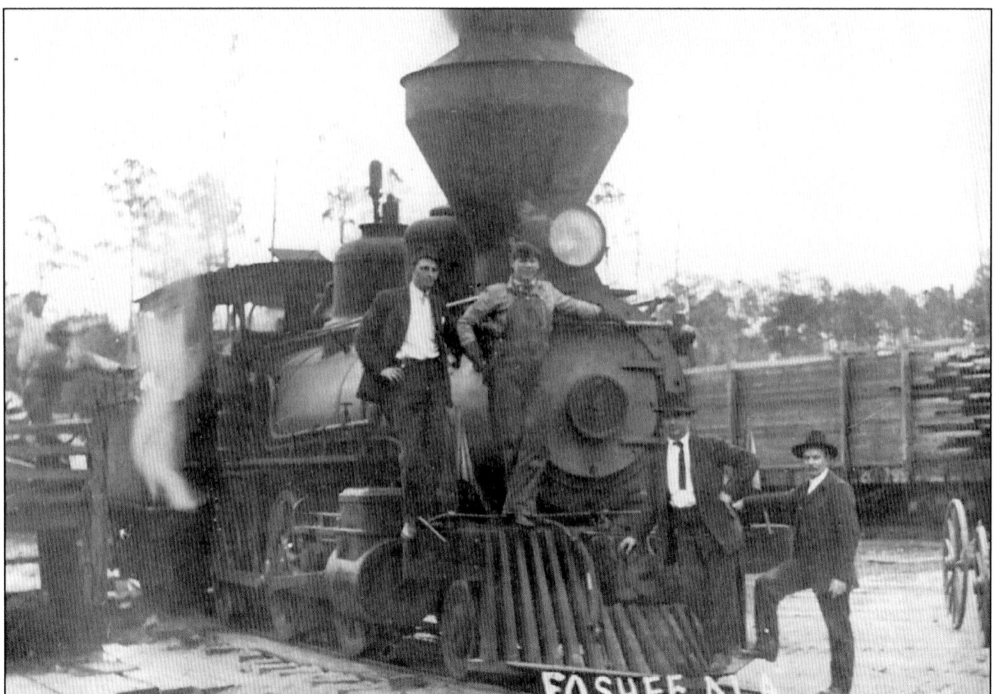

Engine No. 4 makes a stop at Alger-Sullivan's mill in Foshee in 1912. There is no evidence of a mill or a town at Foshee now. (Courtesy of Jerry Fischer.)

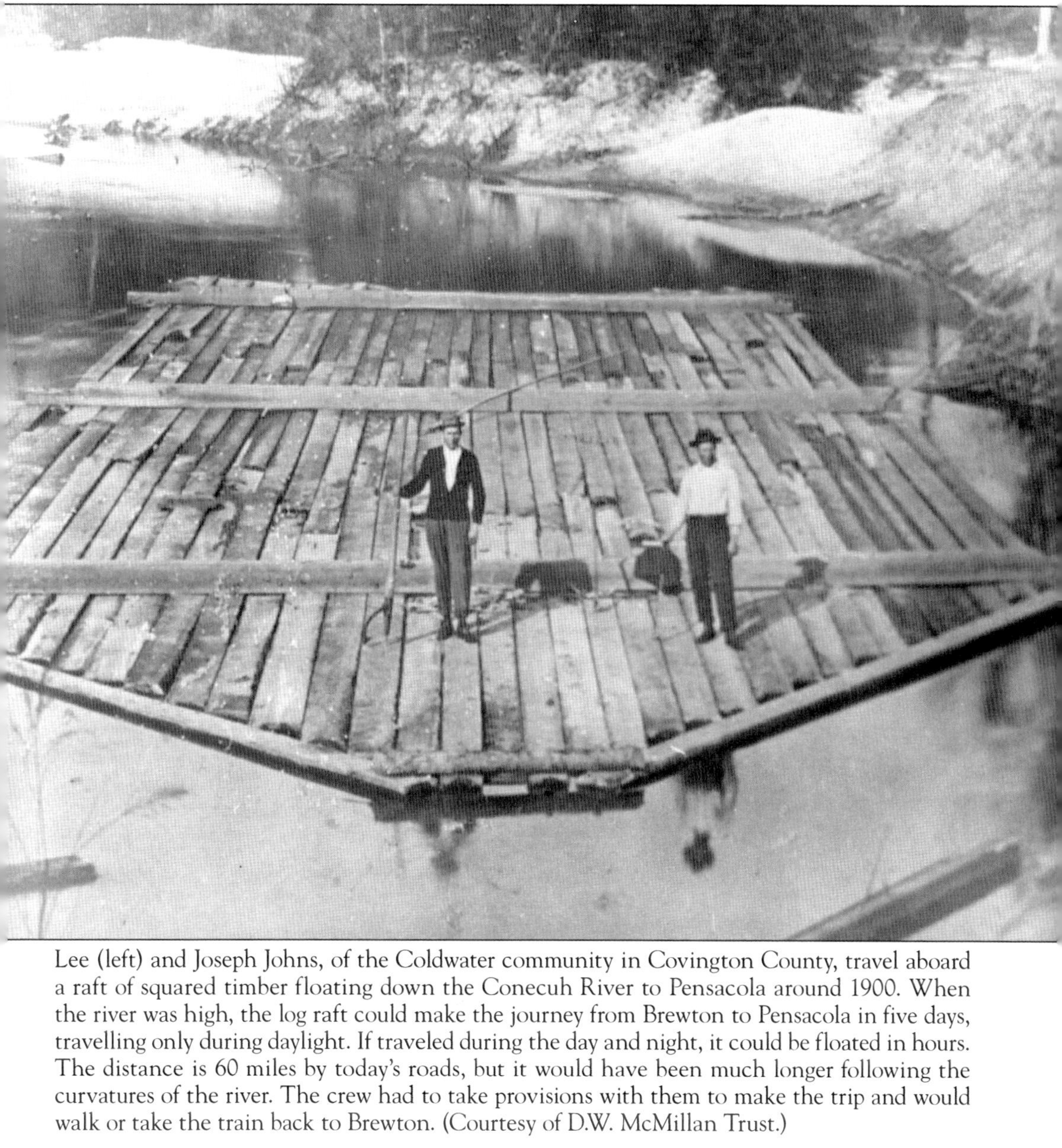

Lee (left) and Joseph Johns, of the Coldwater community in Covington County, travel aboard a raft of squared timber floating down the Conecuh River to Pensacola around 1900. When the river was high, the log raft could make the journey from Brewton to Pensacola in five days, travelling only during daylight. If traveled during the day and night, it could be floated in hours. The distance is 60 miles by today's roads, but it would have been much longer following the curvatures of the river. The crew had to take provisions with them to make the trip and would walk or take the train back to Brewton. (Courtesy of D.W. McMillan Trust.)

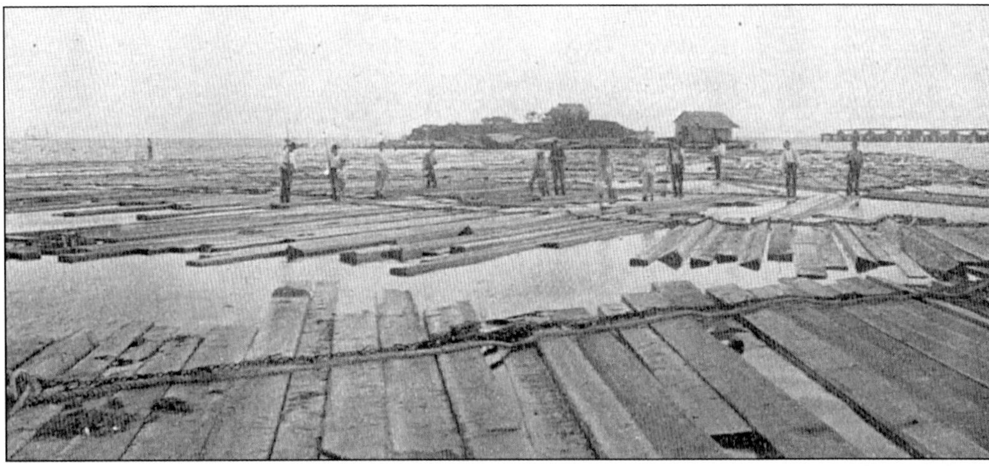

Most of the timber that was cut in the early days of logging in Escambia County floated down the river to Pensacola, Florida, for export. The trees were cut into square timber of 12 inches by 12 inches or so, loaded on ships, and exported to Europe. There was not much of a local market for timber or lumber from the late 1800s to 1920, so the need to export was great. (Courtesy of Jerry Fischer.)

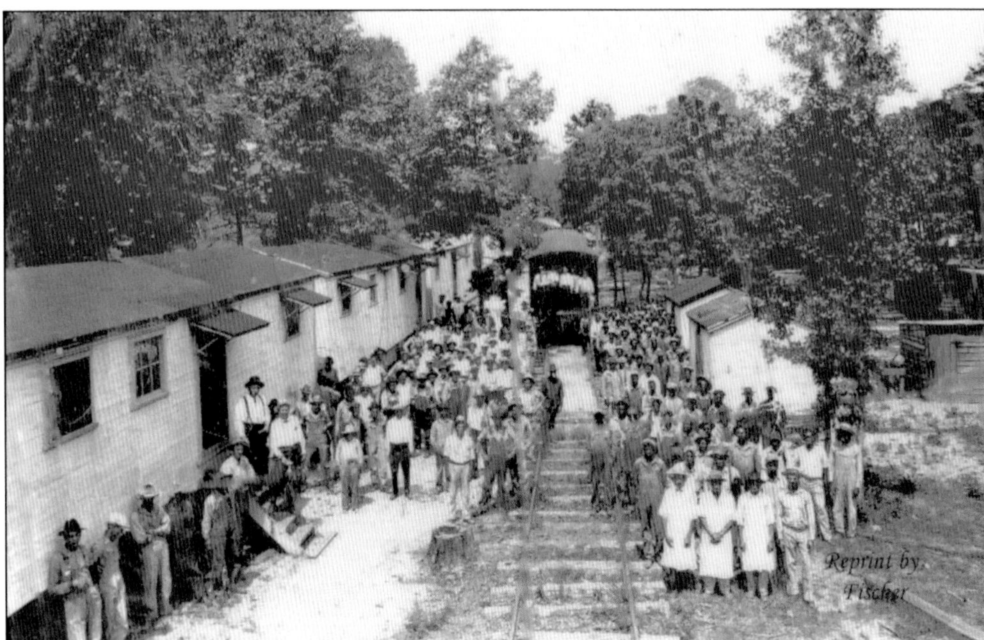

Most of the people worked at Alger-Sullivan Lumber Company's Camp 8 from Sunday evening to Friday evening. The companies would have to provide meals, grow what produce they could in gardens, take care of the animals, and maintain law and order as best they could. (Courtesy of Jerry Fischer.)

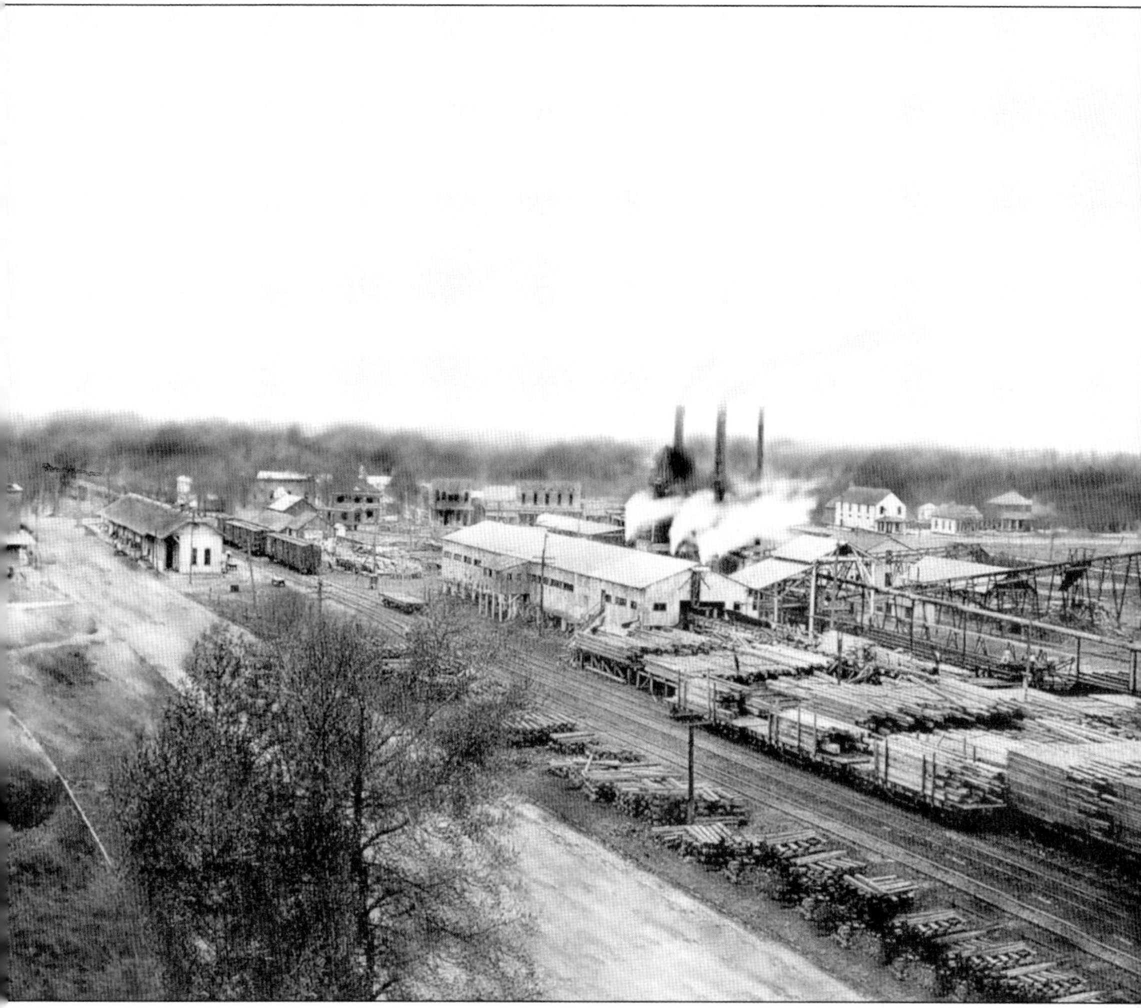

This view of Carney Mill was taken from the top of the water tower in the early 1900s. The timber industry attracted people from many surrounding counties. The Louisville & Nashville Railroad (L&N) train station (center left) handled both freight and passenger service. The nearby commissary at the intersection of Highways 31 and 21 not only fed local workers but also held a regular "chicken catch" contest. A $5 bill was tied to a chicken, and the bird was let go outside of the commissary; whoever could catch the chicken kept it and the $5. (Courtesy of John Webb.)

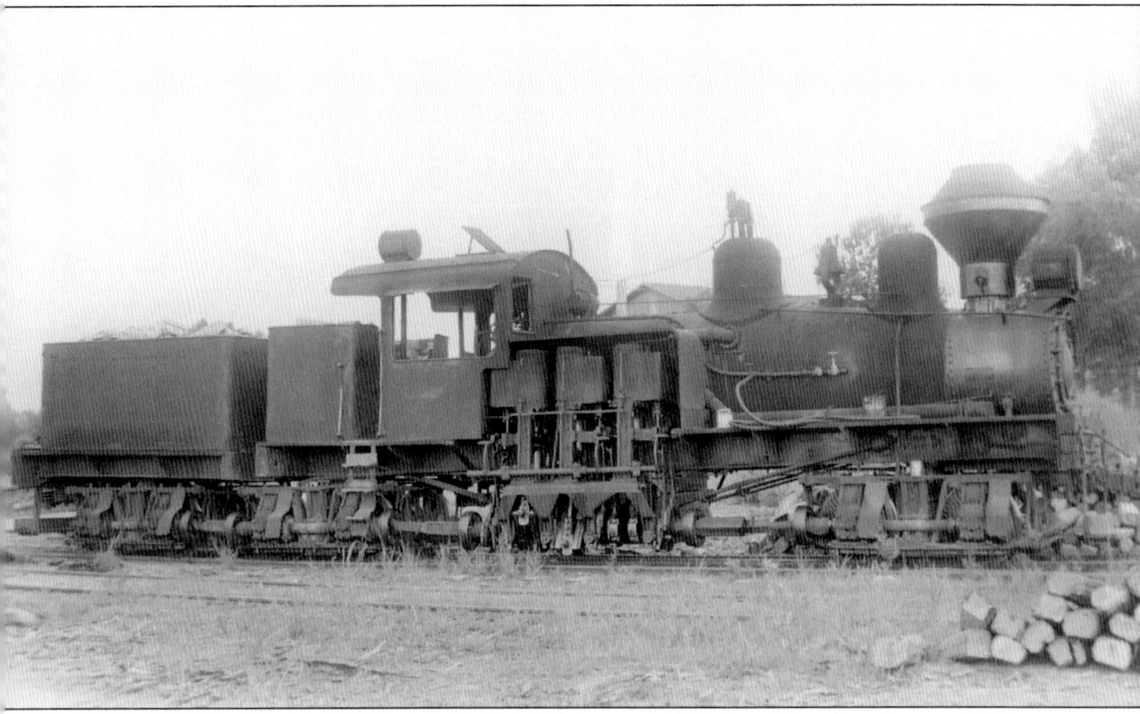

Swift-Hunter Lumber Company No. 60 was a heavy, wood-burning Shay locomotive. Swift-Hunter started its operation in Atmore in 1923 and continues to operate today as Swift Lumber, Inc. The company has the distinction of having the last wood-burning locomotive to operate in Alabama; it was taken off the rails in October 1956. (Courtesy of Thomas Lawson Jr. and C.W. Witbeck.)

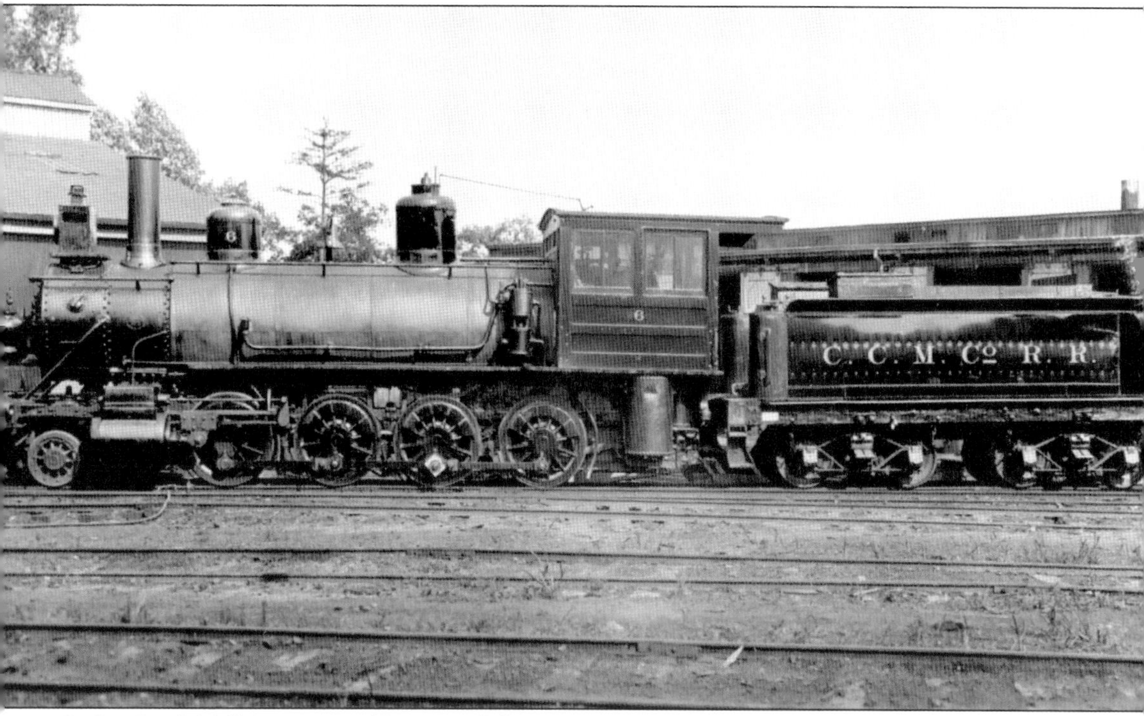

Cedar Creek Mill Company of Brewton did not start using rail to move logs until 1905, but when it did it started with three locomotives, 36 log cars, and 28 miles of track. By 1911, the company had four locomotives, 40 log cars, and 42 miles of track. (Courtesy of Thomas Lawson Jr. and C.W. Witbeck.)

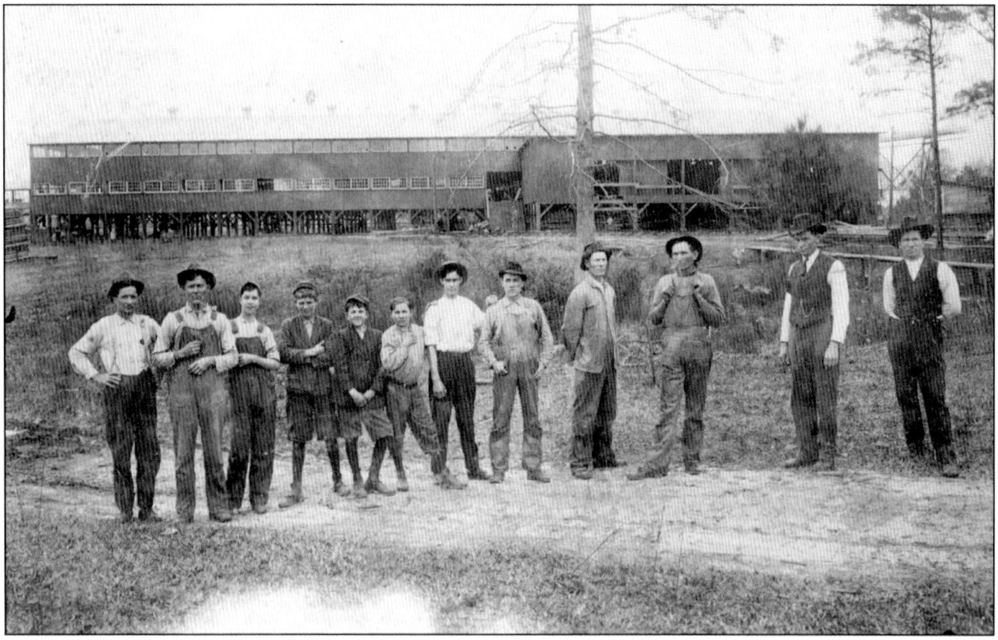

Part of the workforce at the T.R. Miller sash, door, and blind factory takes a break to pose for a photograph. Once the demand for large quantities of wood dried up in Europe, the mills had to make some changes to survive and focused on making more products for domestic use. This division operated from 1903 to 1928. (Courtesy of D.W. McMillan Trust.)

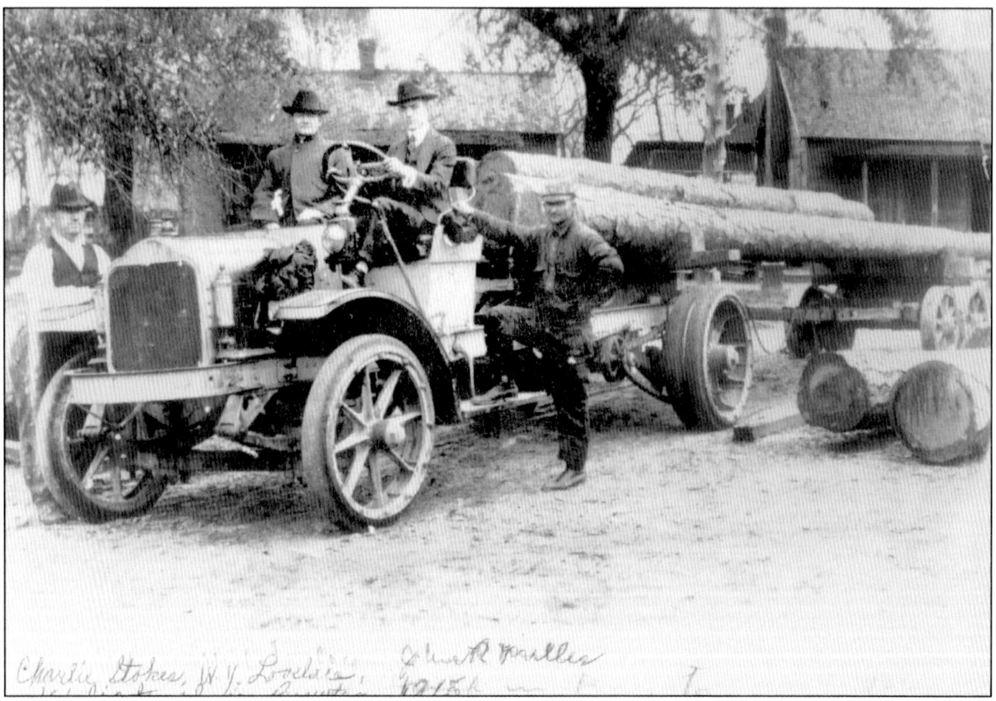

Charlie Stokes (left), W.Y. Lovelace (center), and John R. Miller drive the first logging truck in Brewton in 1915. It was owned and operated by the T.R. Miller Mill Company. Slowly but surely, methods of moving the heavy loads of logs to the mills changed. (Courtesy of D.W. McMillan Trust.)

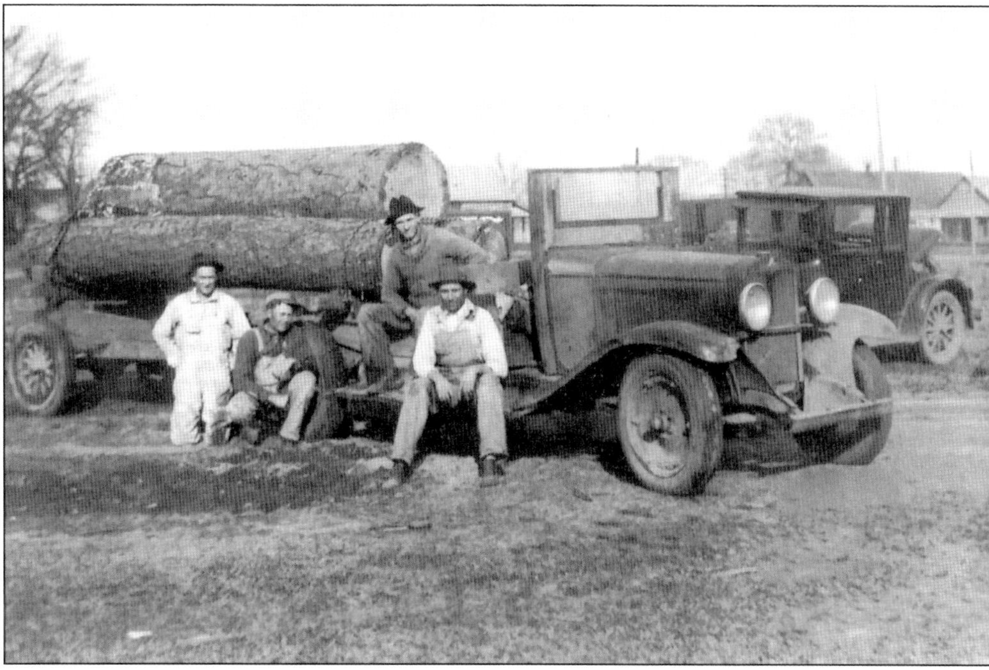

Bondurant family members pose with Bondurant sawmill's first log truck in the Fannie area just north of Flomaton. The Bondurant family has had a continuous operation in the timber business since this picture was taken. (Courtesy of Jerry Fischer.)

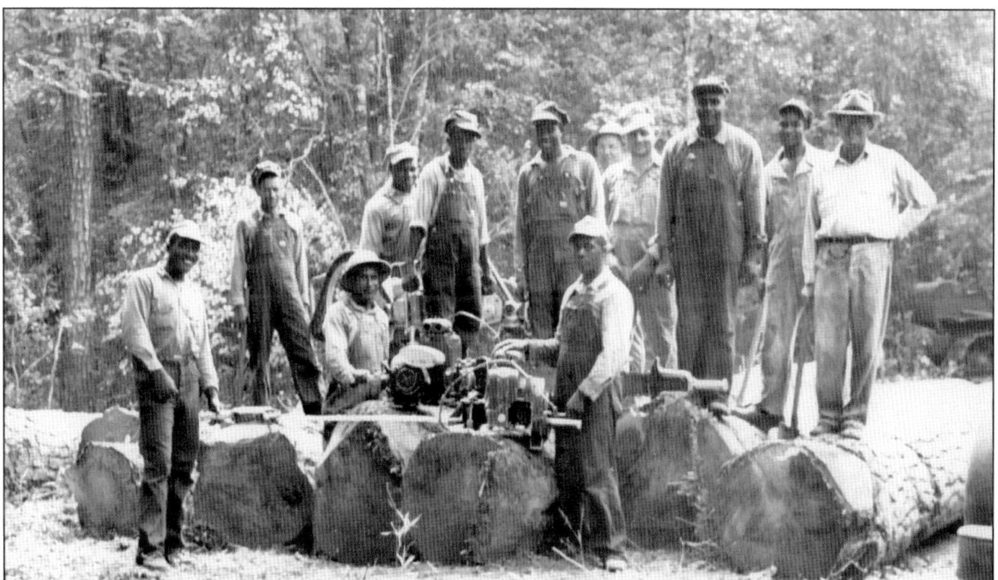

The use of chainsaws became common after World War II. This 1951 photograph of a T.R. Miller crew shows one of the early large chainsaws. The men in the picture are, from left to right, Herman Williams, Elisha Likely, John Mix, Lige Culliver, Kyler Benjamin, David Benjamin, Henry Rankin, Chesley Caro, Claude Rains, Angelo Culliver, Claude Rankin, and Jack Lucas. (Courtesy of D.W. McMillan Trust.)

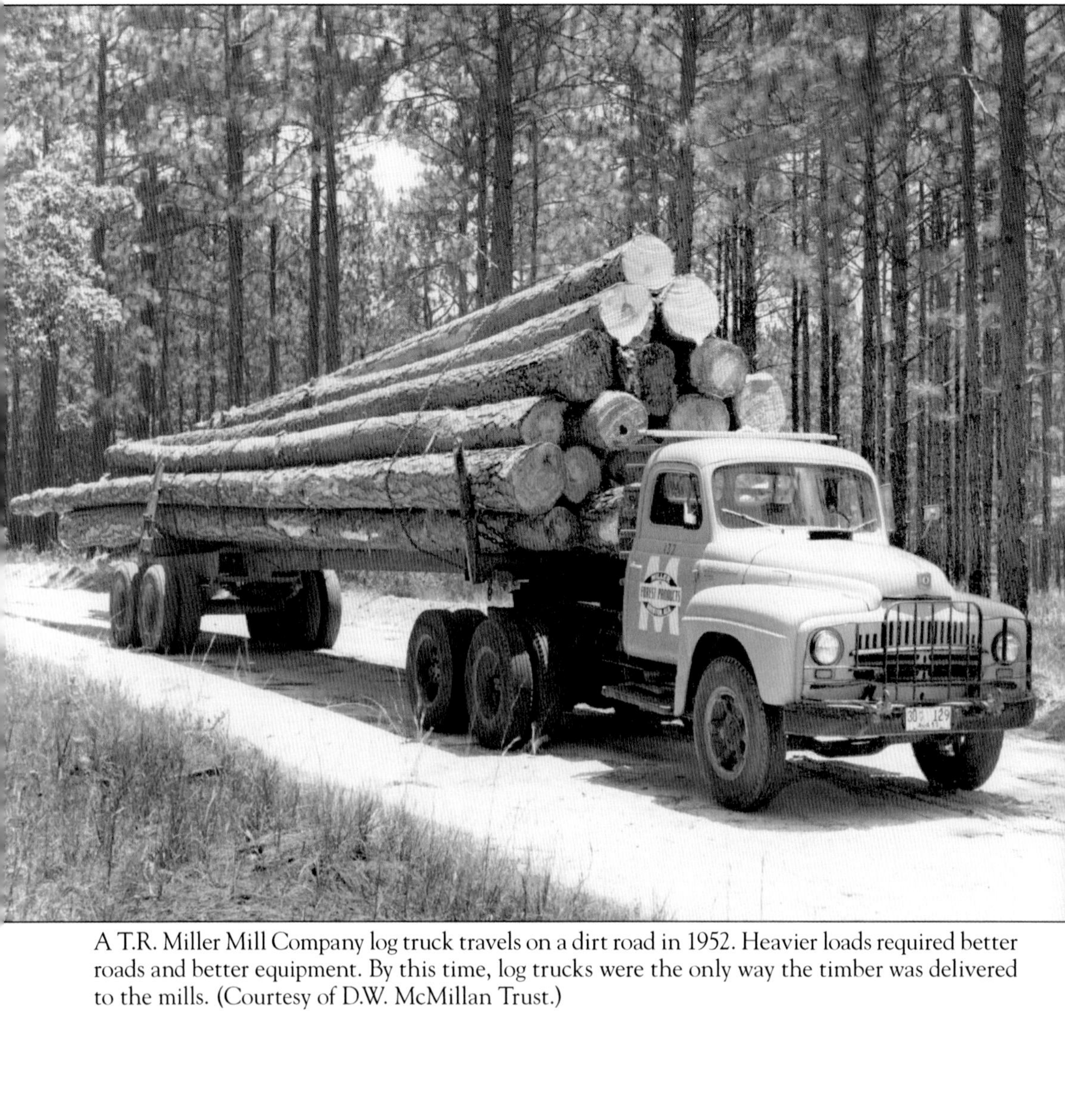

A T.R. Miller Mill Company log truck travels on a dirt road in 1952. Heavier loads required better roads and better equipment. By this time, log trucks were the only way the timber was delivered to the mills. (Courtesy of D.W. McMillan Trust.)

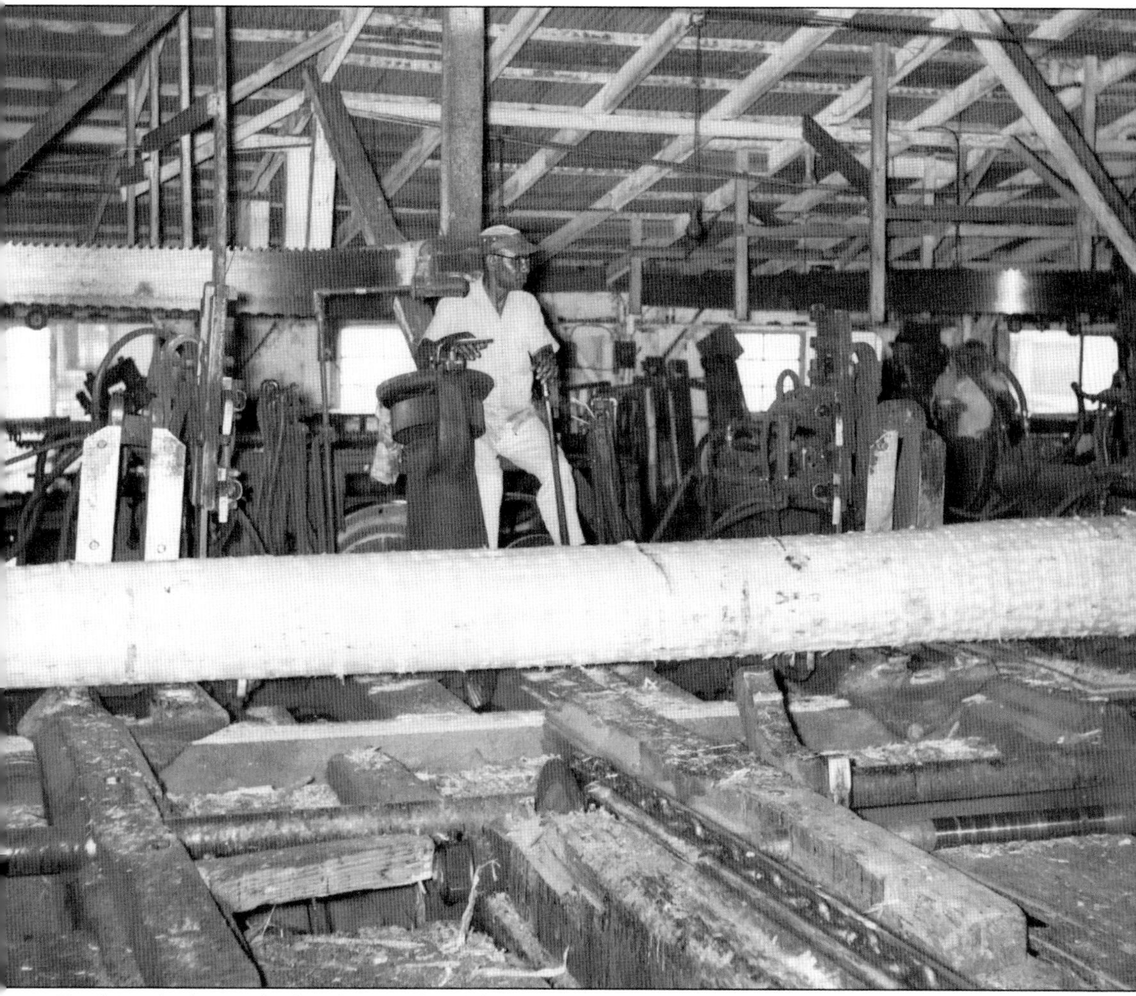

Fred Sanford cuts the last log on the old carriage in the T.R. Miller Mill Company sawmill in Brewton on July 1, 1958, at 5:00 p.m. on a Tuesday afternoon. Sanford had been a block setter on this carriage for 36 years. (Courtesy of D.W. McMillan Trust.)

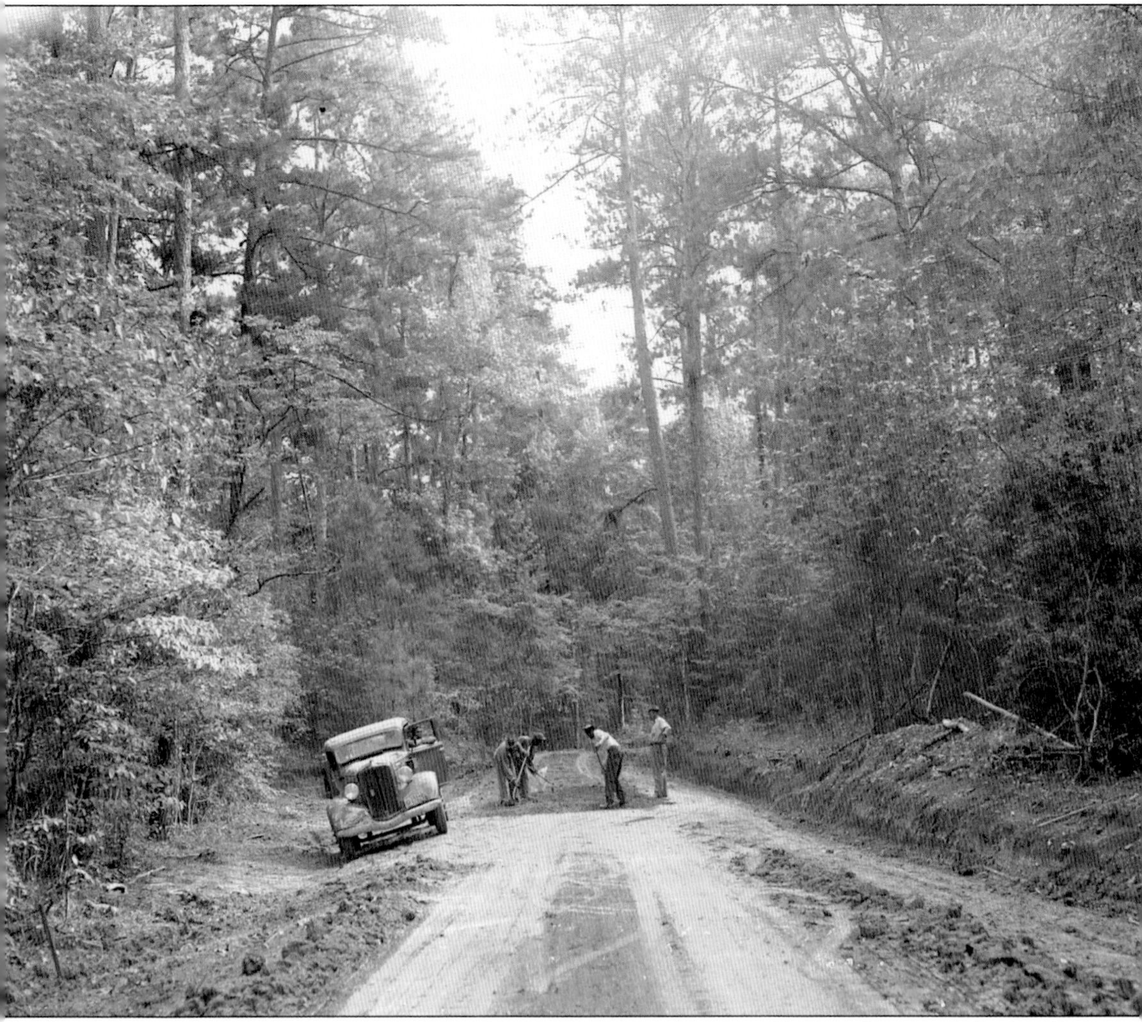

After the virgin forest was cut from the 1870s to 1920s, many mills shut down and the land lay idle; however, some landowners replanted or encouraged natural regeneration by leaving seed trees in the forest. This stand of timber in the 1950s is a good example of the second generation of trees, with an estimated age of at least 60 years. (Courtesy of D.W. McMillan Trust.)

Three

Sports, Recreation, and Entertainment

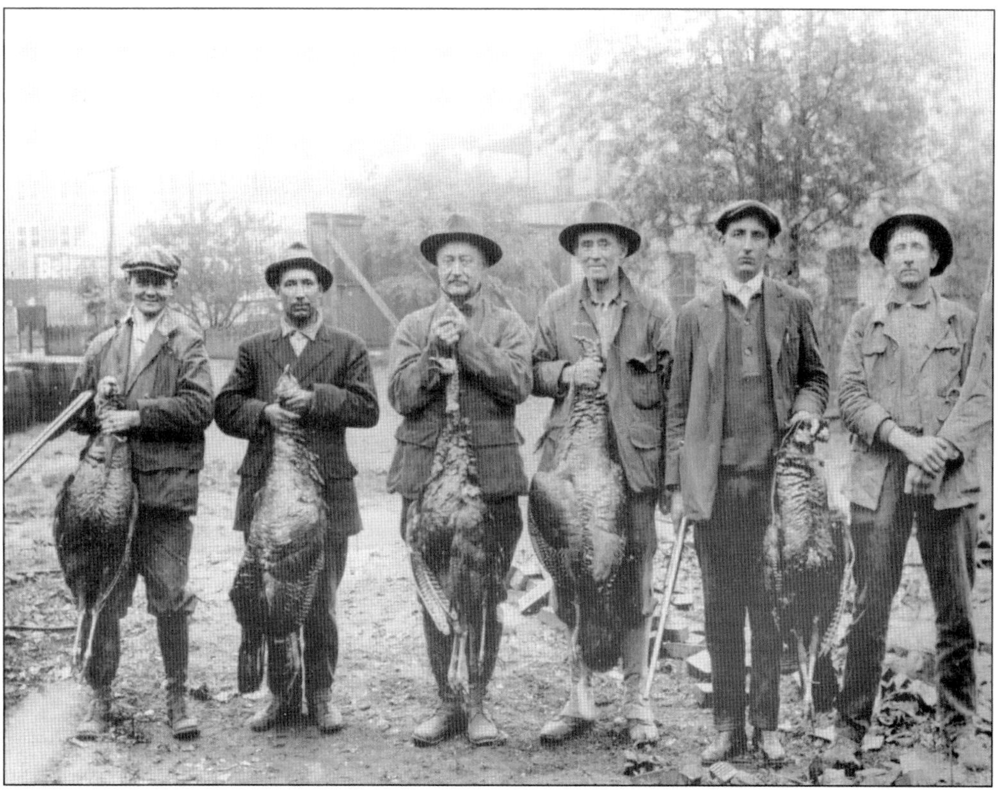

The hunting party of a successful 1920 turkey hunt shows off its game in downtown Brewton. Wild turkeys thrive in the area's piney woods and creek bottoms. This area is known for its hunting opportunities, including whitetail deer, quail, dove, and raccoon. From left to right are Ed Leigh McMillan, Young Padgette, Judge R.T. Ervin (of Mobile), V.C. Griffins, Dunk Liles, and Bill Reeves. (Courtesy of D.W. McMillan Trust.)

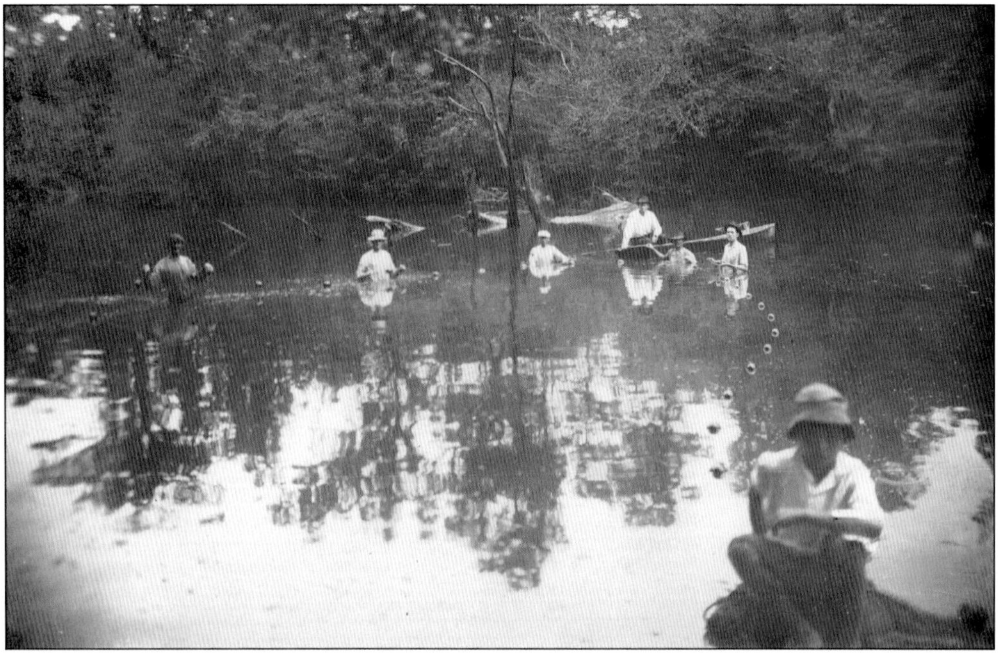

These men and one young person are putting out a net in one of the many natural lakes in the Pollard swamp area in the Conecuh River bottom around 1910. Hunting and fishing have always been part of the culture in this area. (Courtesy of D.W. McMillan Trust.)

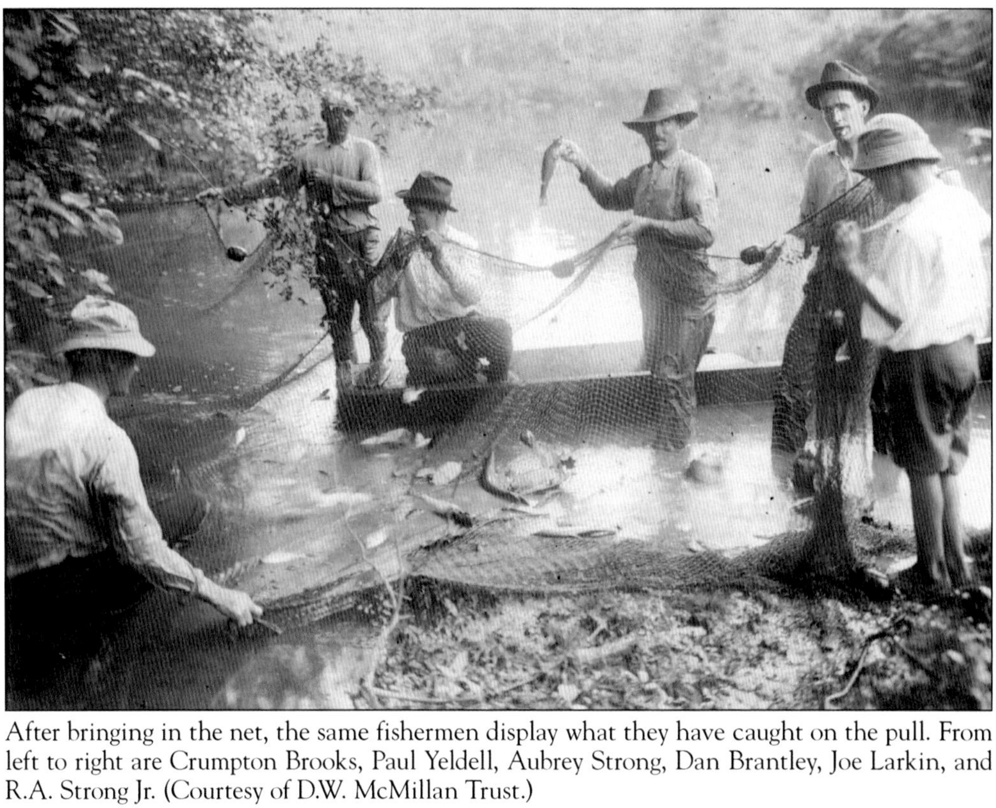

After bringing in the net, the same fishermen display what they have caught on the pull. From left to right are Crumpton Brooks, Paul Yeldell, Aubrey Strong, Dan Brantley, Joe Larkin, and R.A. Strong Jr. (Courtesy of D.W. McMillan Trust.)

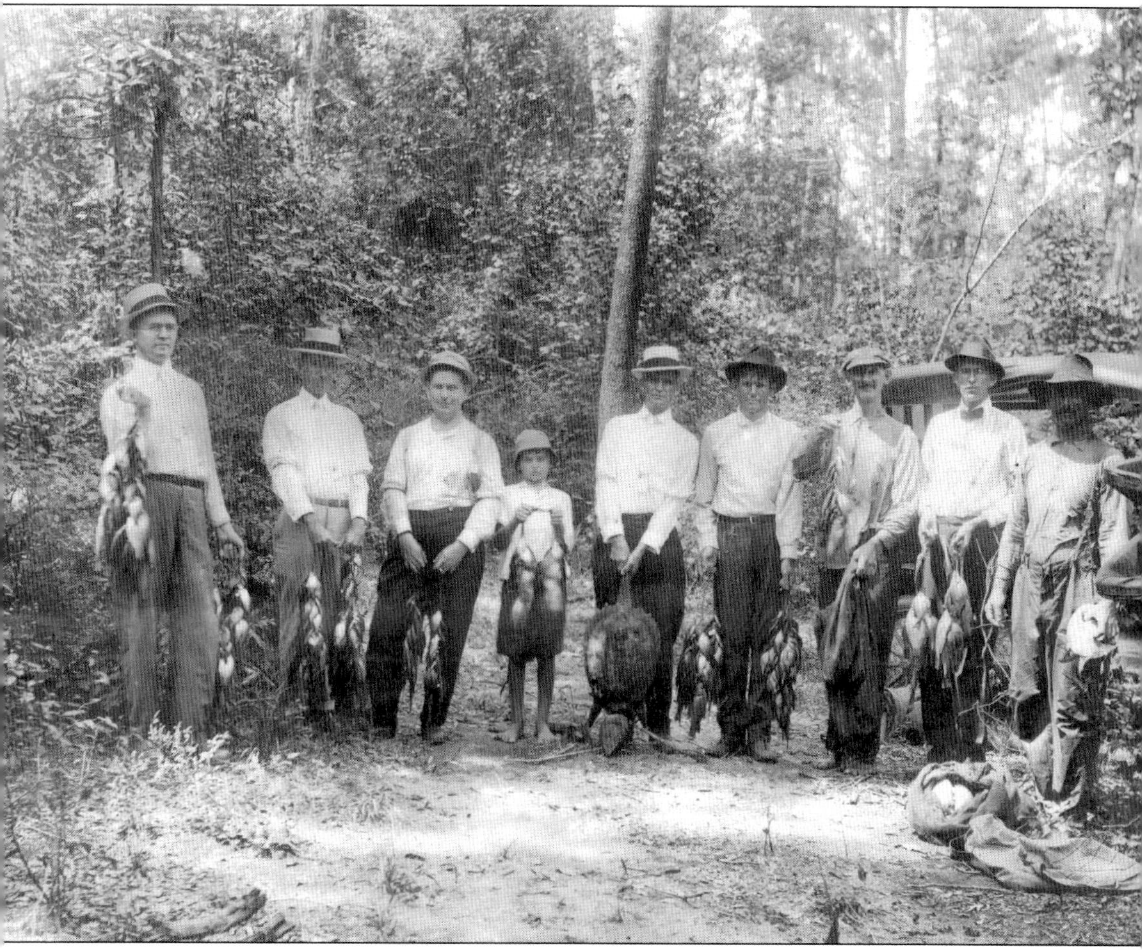

At the end of day and after many pulls of the net, the same group of fishermen shows off its catch. The man in the middle is holding a large snapping turtle. The men (and boy) are, from left to right, C.C. Brooks, Young Padgett, Clarence Reid, R.A. Strong Jr., E.P. Yeldell, R.A. Strong Sr., George Mantle, Joe Larkin, and Dan Brantley. (Courtesy of D.W. McMillan Trust.)

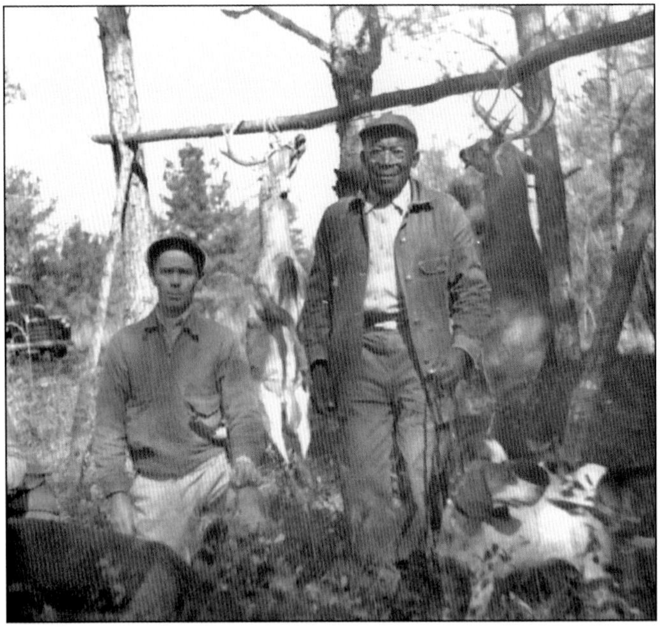

Julius Taylor (left) and Luke Fowler pose with a couple of nice bucks killed near the Sepulga River in 1951. This woodsy area is in a rural part of the county near the town of Brooklyn. (Courtesy of D.W. McMillan Trust.)

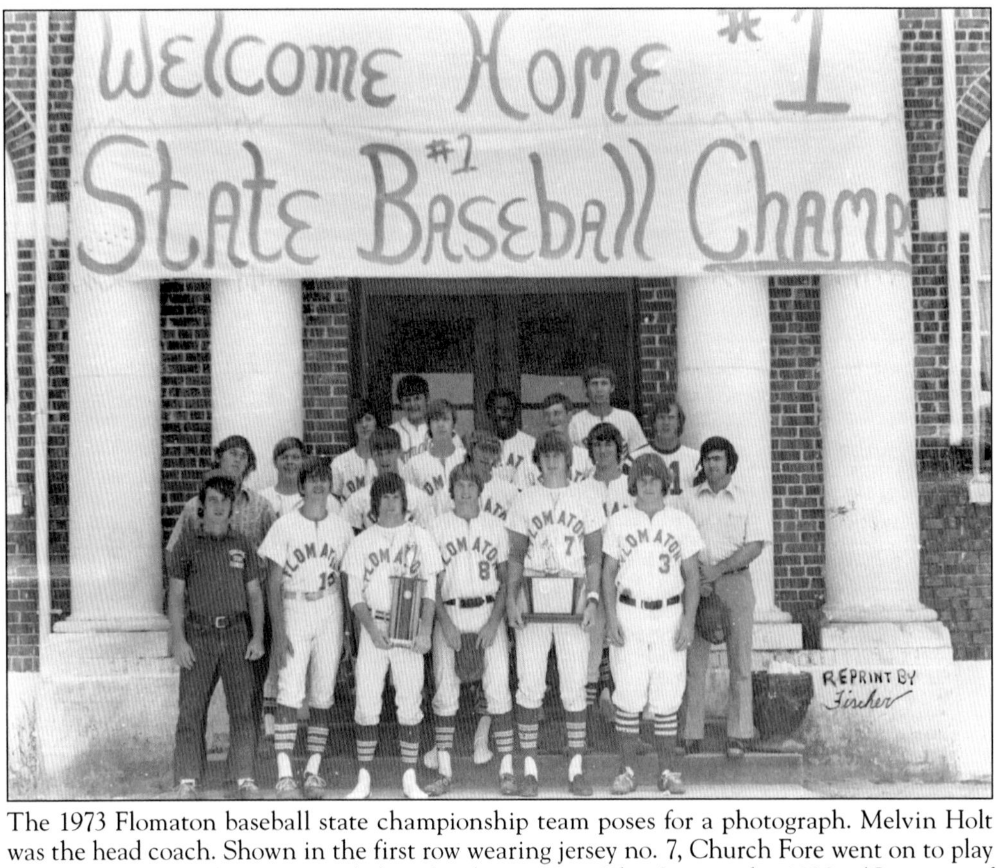

The 1973 Flomaton baseball state championship team poses for a photograph. Melvin Holt was the head coach. Shown in the first row wearing jersey no. 7, Church Fore went on to play at the University of Georgia and then for the Toronto Blue Jays in the 1980s. (Courtesy of Jerry Fischer.)

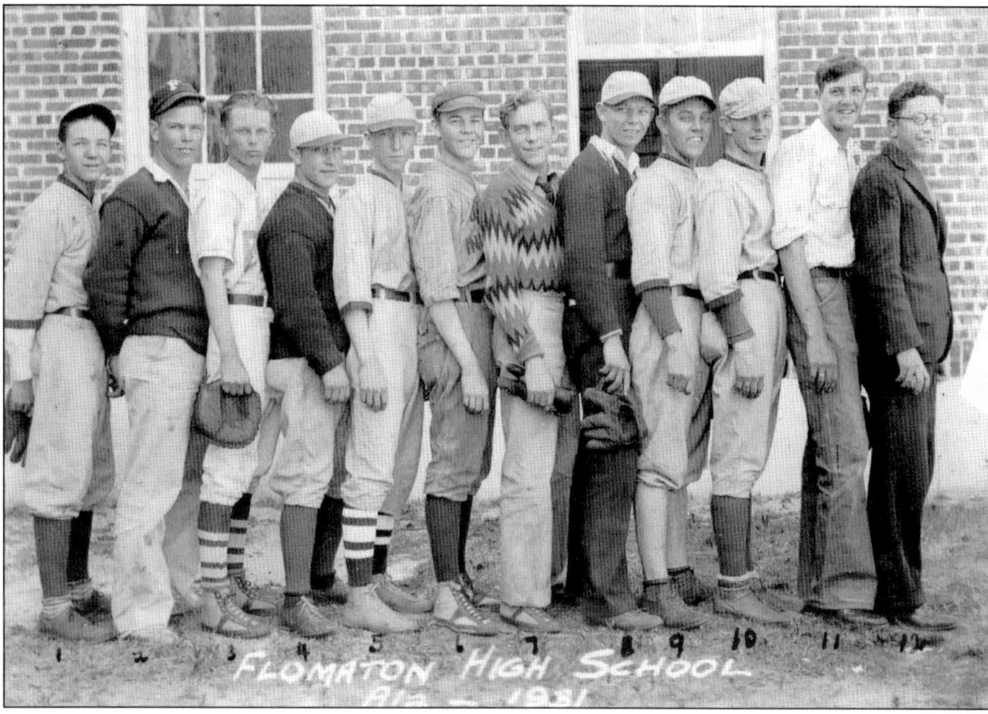

Members of the Flomaton High School 1931 baseball team line up for a photograph. They won eight games and lost two; Sammy Jackson is fourth from the right. Flomaton High School has always had strong teams in baseball, football, and basketball with players, like Lefty Adams, "Big" Ed Morris, Chuck Fore, and Buck Showalter from nearby Century, Florida. (Courtesy of Jerry Fischer.)

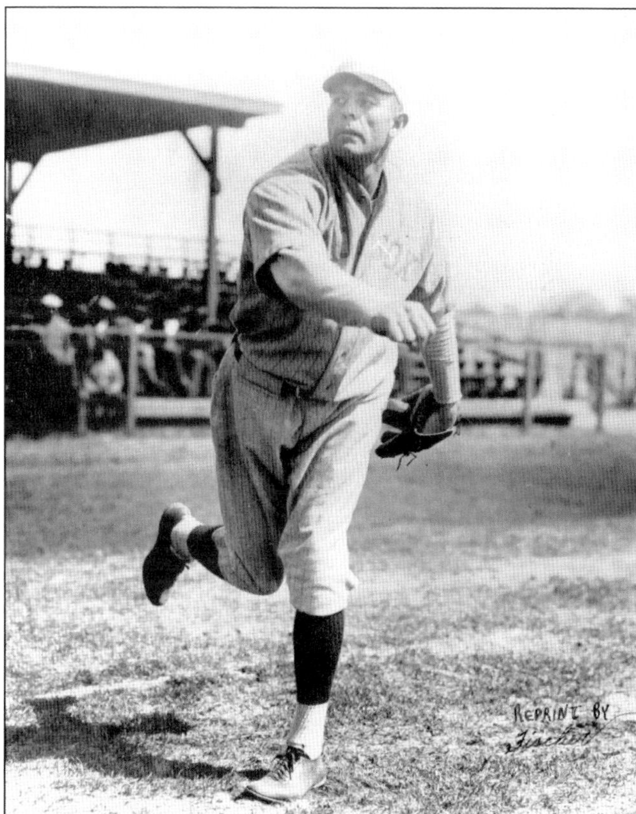

Big Ed Morris, of Foshee, was one of the most famous athletes from Escambia County. He pitched for the Chicago Cubs and the Boston Red Sox between 1922 and 1931. By 1932, he was developing into quite a player. That spring, he came home for a visit and was killed at a party held in his honor. He died at the age of 32. (Courtesy of Jerry Fischer.)

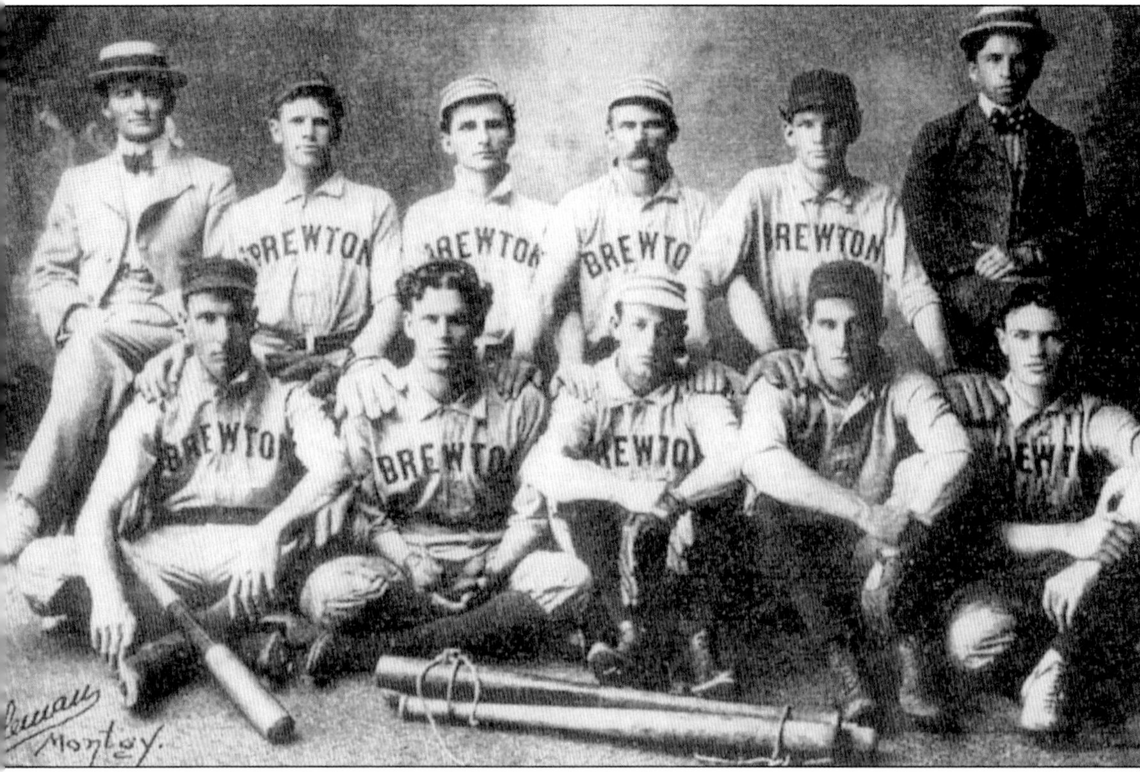

The 1901 Brewton Baseball Club lines up for a photograph, one that has become a local classic. The team traveled and played other towns in south Alabama. In 1914, they were champions of south Alabama. William Yancey Lovelace (back left) was the manager. (Courtesy of Escambia County Historical Society.)

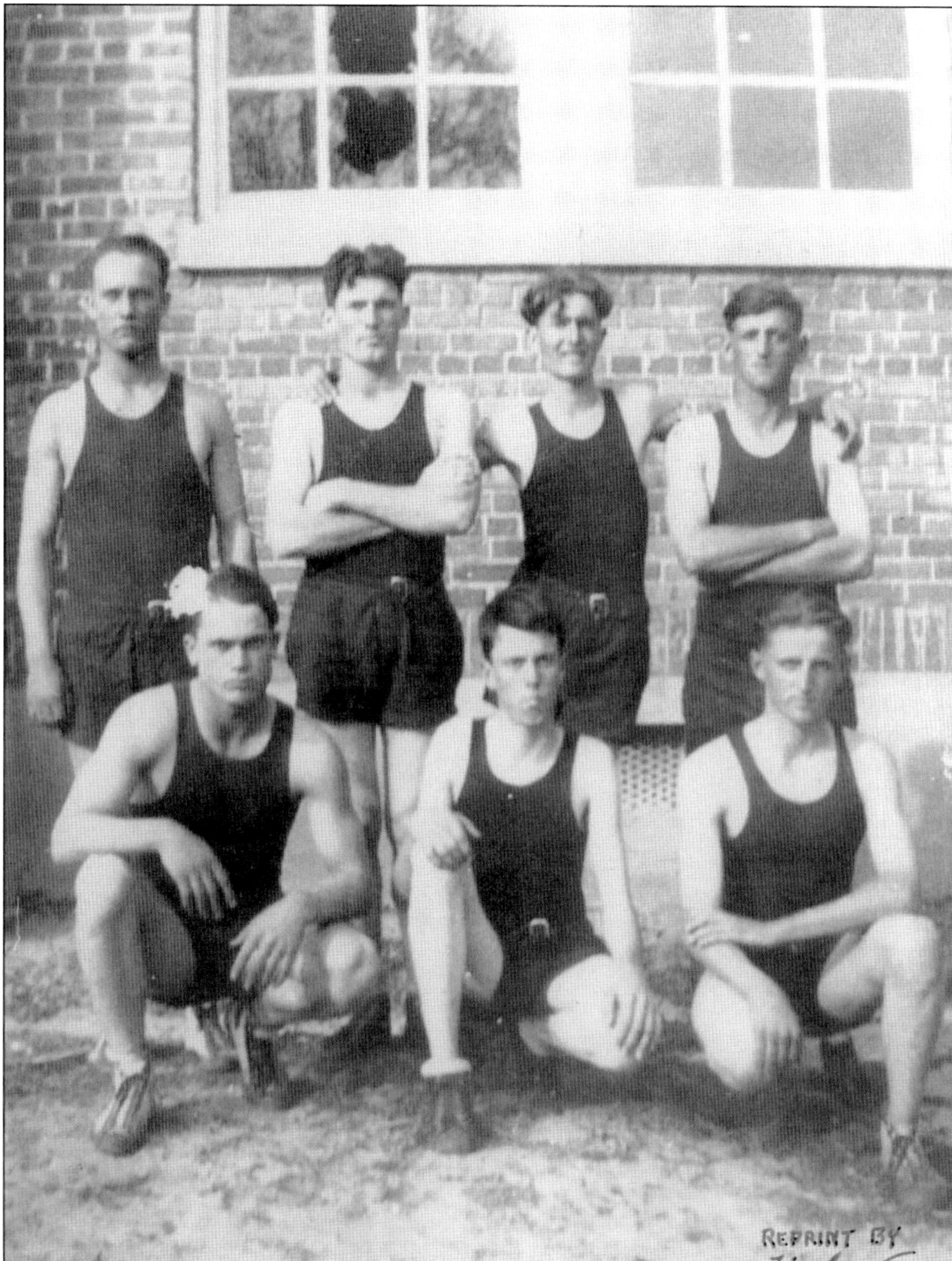

The McCall School basketball team was the state champion of the 1929–1930 season. Coach Buck Jones (not shown) has the distinction of winning 54 straight games around this time in his career. These players include, from left to right, (first row) M. Brantley, Norman Tedder, and B. Crossley; (second row) C. Bradley, Ray Tedder, C. Spears, and I. Crossley. (Courtesy of Jerry Fischer.)

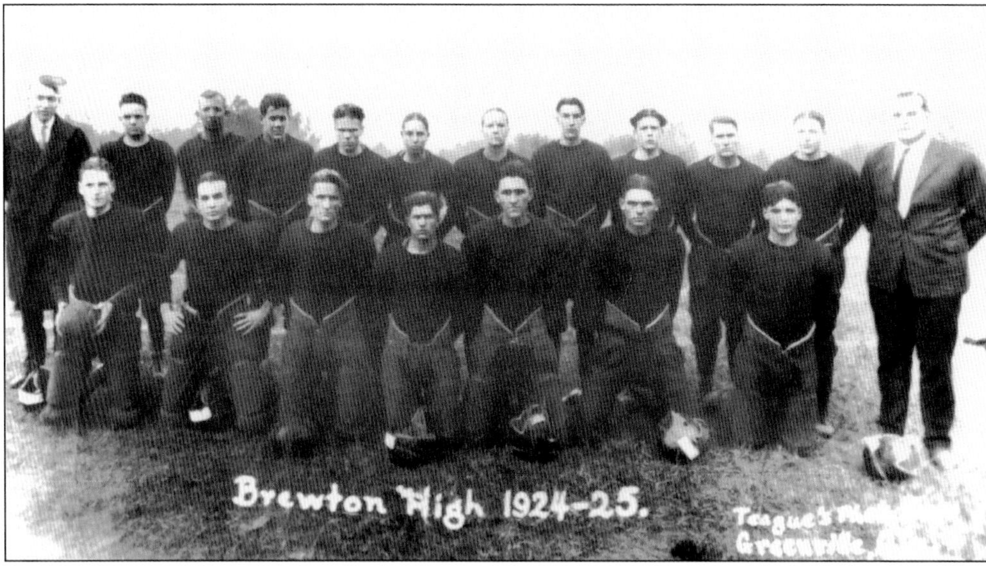

The Brewton High football team gathers for a picture in 1925 during the days of leather helmets and thin pads. From left to right are (first row) Durwood McPhaul, Dick Jordan, Houston Odom, Tich Bedgood, Edwin Malone, Gus McPhaul, and R.A. Strong; (second row) coach Chapman Curry, Curtis Finlay, Coleman Fuqua, Jim Lovelace, Ernest Hand, Boots Miller, Edgar Red Moor, Lawrence Holiday, Joe Cooper, J.C. Temple, Ned Grice, and coach Ike Rodgers. (Courtesy of Escambia County Historical Society.)

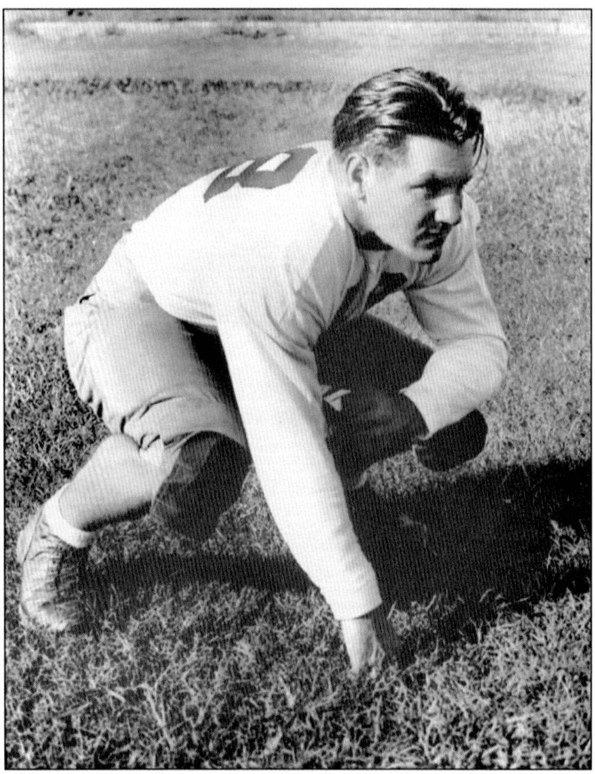

Art "Tarzan" White, a native of Atmore who got his name from swinging on vines over creeks in the area, began his long football career at Escambia County High School. At age 13, he was already five feet, eight inches tall and 213 pounds. He played for the University of Alabama on the 1934 national championship team. After college, Tarzan was a first-round draft pick of the New York Giants, and he won a world championship with them in 1938. During his six-year career, he also played for the Chicago Cardinals and was a member of the first Pro Bowl team. (Courtesy of the University of Alabama.)

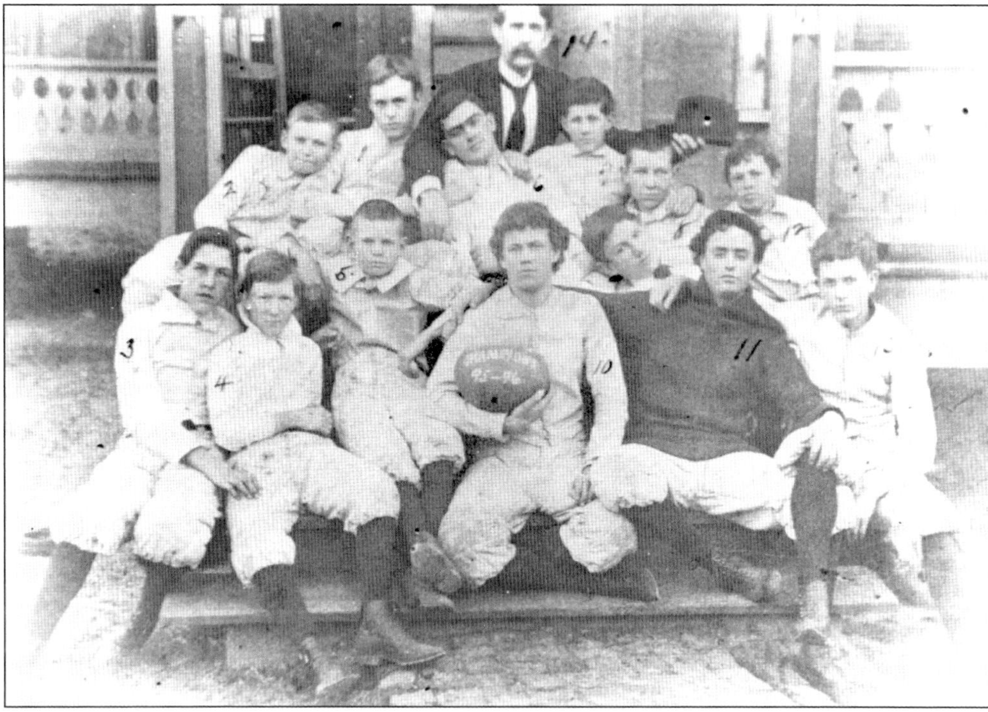

Sports were important in the early days. The Brewton Collegiate Institute football team was the south Alabama champion of the 1895–1896 school year. From left to right are (first row) Morse Brooks (3), Oscar Colley (4), R.A. Strong (5), H.C. Rankin (10), John James (9), S.S. Foshee (11), and R.C. Granberry (13); (second row) Cleve Henderson (2), D.B. Hayes (1), W. McGowin (6), coach Prof. J.V. Brown (14), W.W. Lovelace (7), Ray Walker (8), and Howard Rankin (12). (Courtesy of D.W. McMillan Trust.)

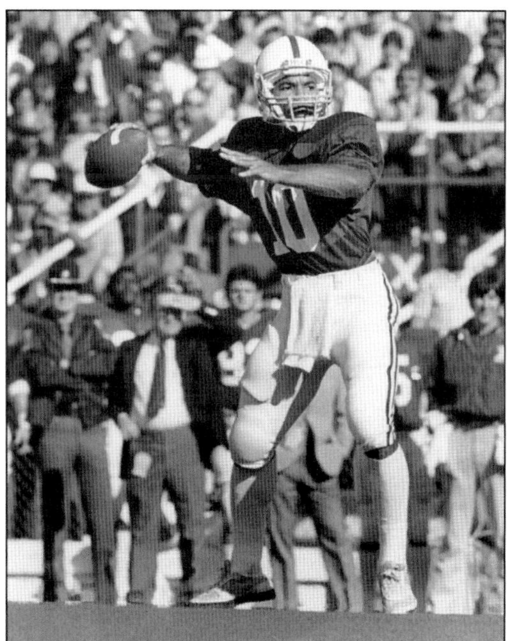

Walter Lewis was a star quarterback for T.R. Miller High School and was signed by the University of Alabama, where he played from 1980 to 1983 as quarterback. He went on to play in the USFL for the Memphis Showboats in 1984 and 1985 and is now enjoying a successful career in business. (Courtesy of the University of Alabama.)

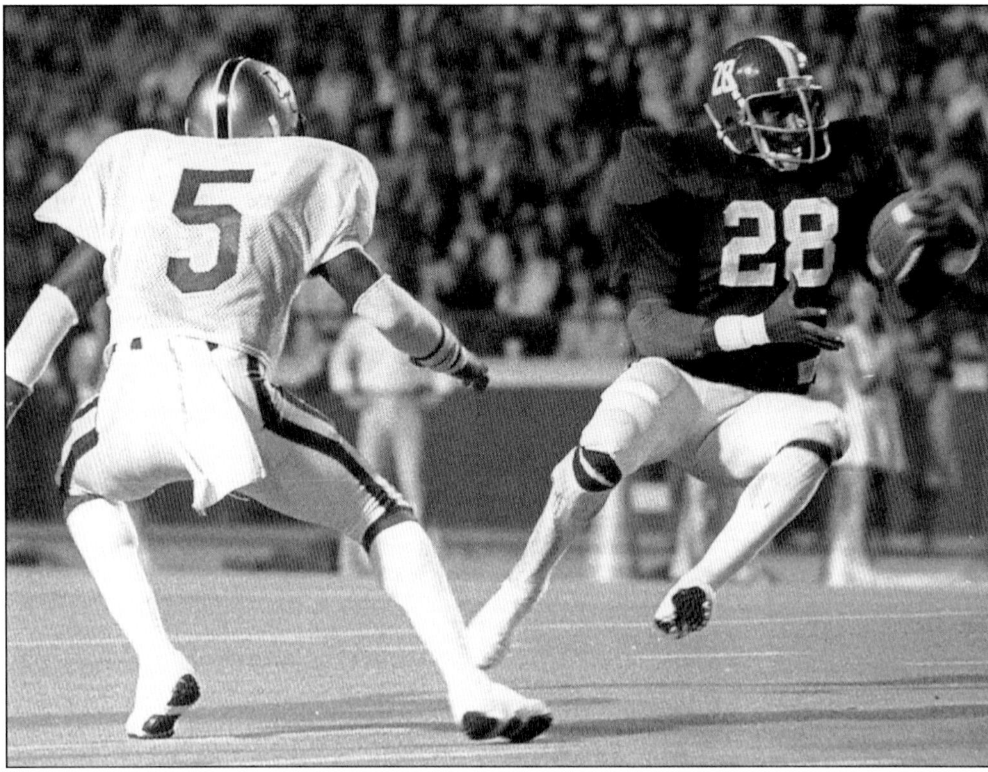

Don McNeal (no. 28) played high school football in Atmore, Alabama, at Escambia High and went on to play at the University of Alabama for Paul "Bear" Bryant on the 1978 and 1979 national championship teams. He was drafted by the Miami Dolphins in 1980 and played there until he retired in 1989. He played in Super Bowls XVII and XIX and was named Dolphins' Player of the Year in both 1982 and 1984. (Courtesy of the University of Alabama.)

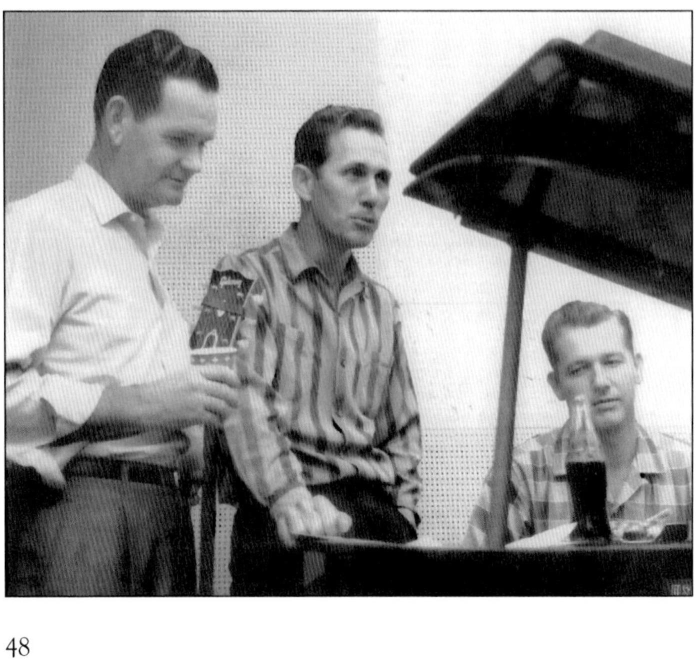

Hank Locklin, who grew up in McLellan, Florida, and later in life became a resident of Brewton, appears beside a piano in 1963 at a recording session at RCA Studio B. Locklin was one of country music's greatest tenors and scored one of the biggest hits of the music's golden era with "Please Help Me, I'm Falling." In 1960, he became a member of the Grand Ole Opry. From left to right are Locklin, Chet Atkins, and Floyd Cramer. (Courtesy of Adam Locklin.)

William Lee Golden and his little sister Lanette Golden Lowery are shown around 1950 when they were singing in Brewton at WEBJ on *The Old Country Church on the Air*, their grandfather's radio show. William Lee went on to become a member of the Oak Ridge Boys, who had over 12 no.-1 hits and 30 top-10 hits. (Courtesy of Brenda Golden.)

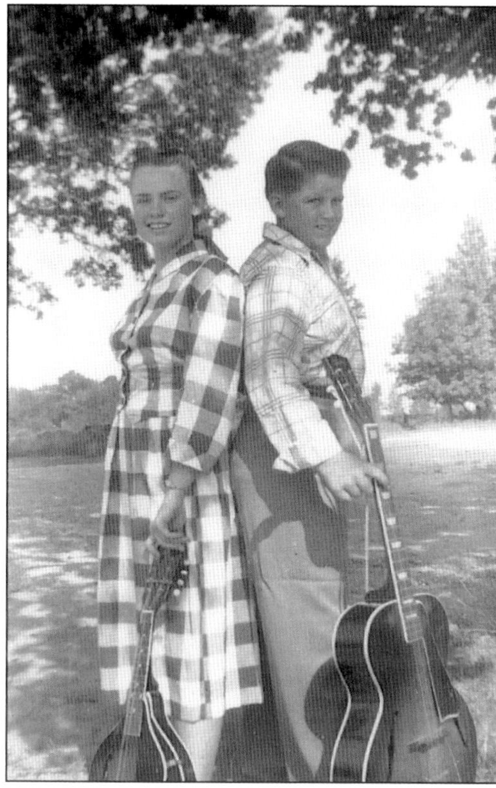

These fine men are being honored at a luncheon for the past presidents of the Brewton Country Club put on by Walter and Elmira Curry. Walter Curry is standing in the center of the picture. He was a longtime employee of the club and wanted to give recognition to the former presidents. Sitting from left to right are C.C. Huxford Jr., W.T. Neal Jr., J.R. Miller Jr., Percy F. Watson, David B. Miller, J.W. Adkisson Sr., W.F. Wilson, J. Pat Hayes, Troy C. Brannan, Ben Kelly Strain, David O. (Mike) Allen, and Casey Jackson. (Courtesy of Yank Lovelace.)

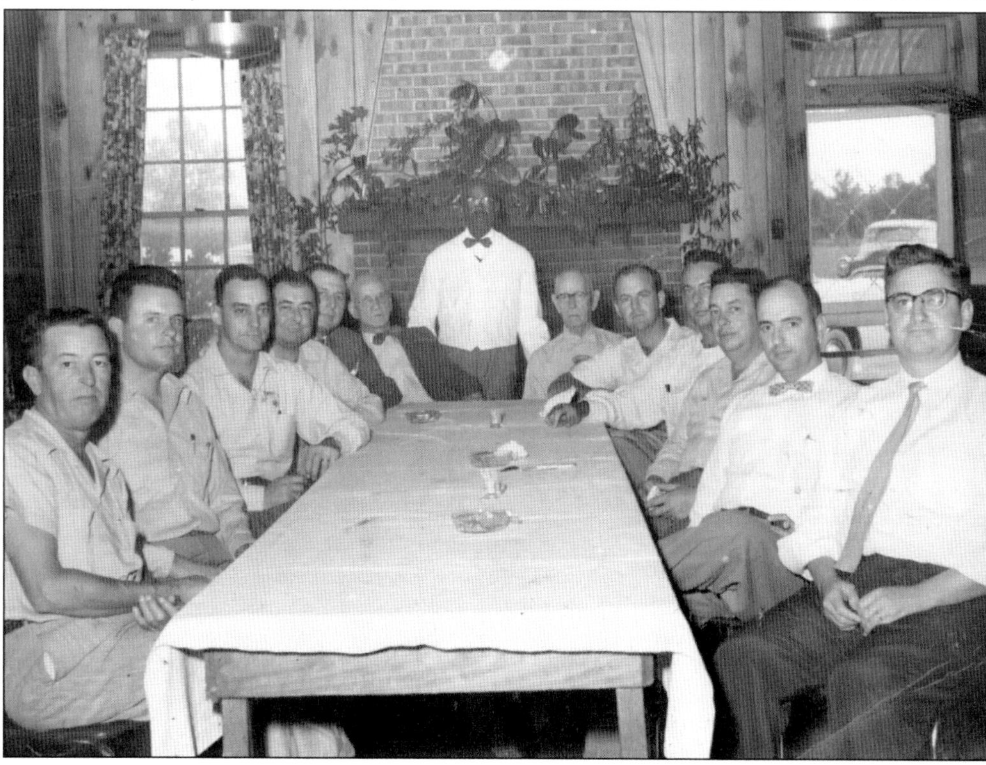

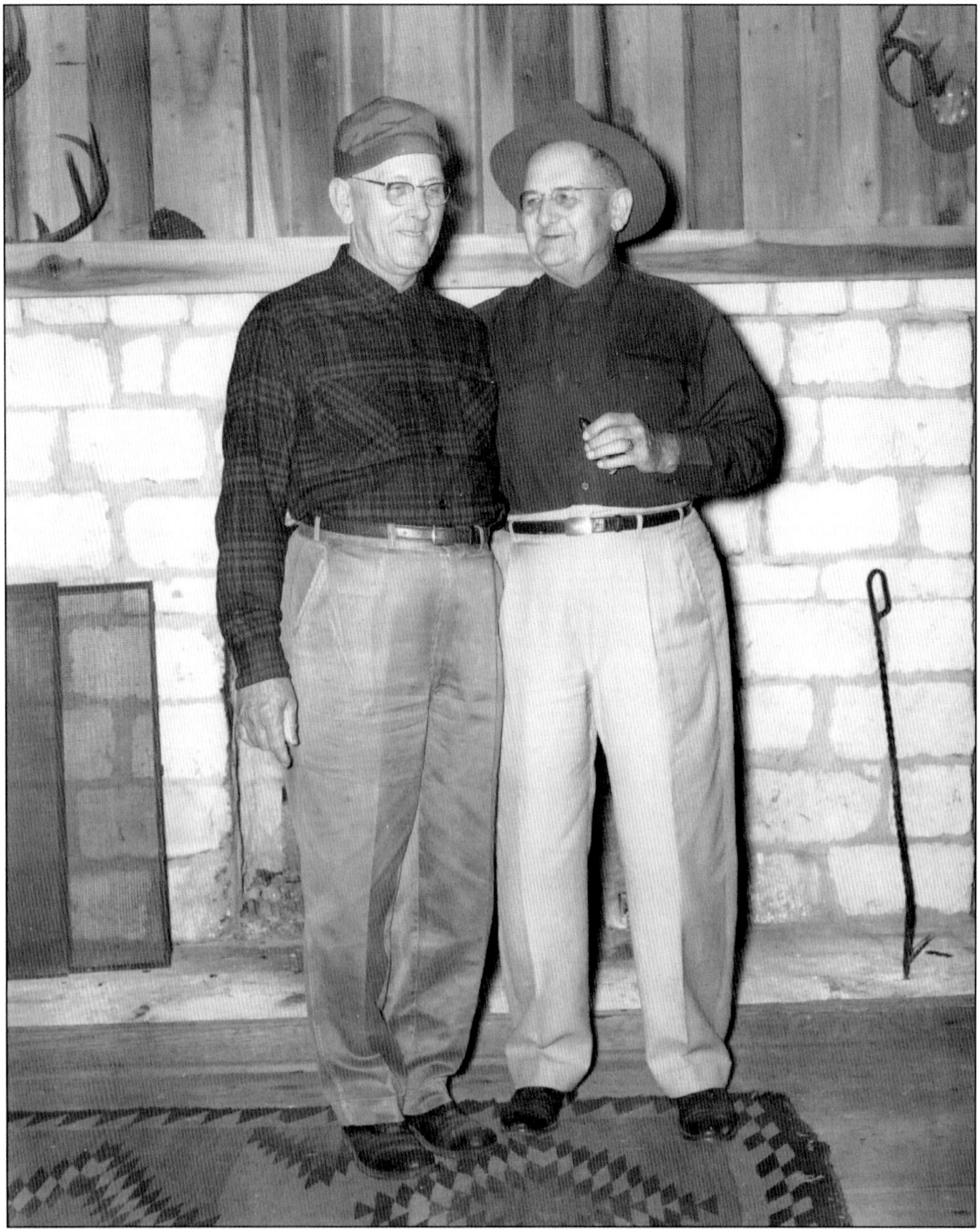

Camp-house camaraderie has always been part of the hunting scene in Escambia County. Besides the hunt, camp houses are used for social events, entertaining friends, and spending family time together. This camp was organized by longtime T.R. Miller Mill Company comptroller Clay Alexander (right). He was hosting a deer hunt for the Birmingham chapter of the Controllers Institute of America in December 1957. (Courtesy of D.W. McMillan Trust.)

Four

DAILY LIFE THROUGH THE YEARS

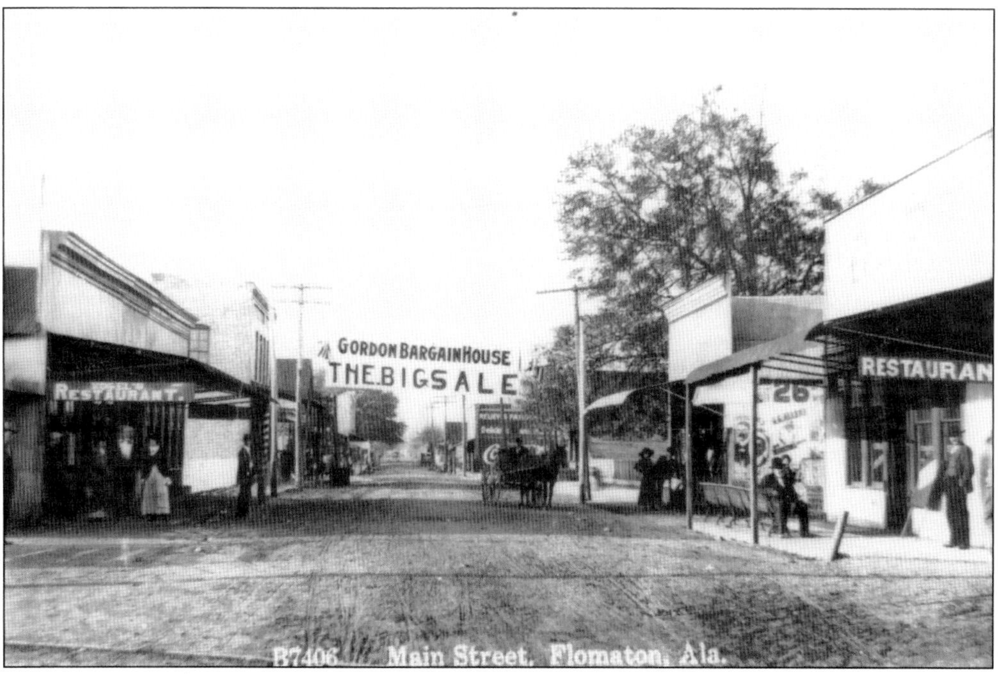

Main Street in Flomaton is shown in the 1920s before paved roads. Looking north, horse and buggy share the road with the early automotive. (Courtesy of Jerry Fischer.)

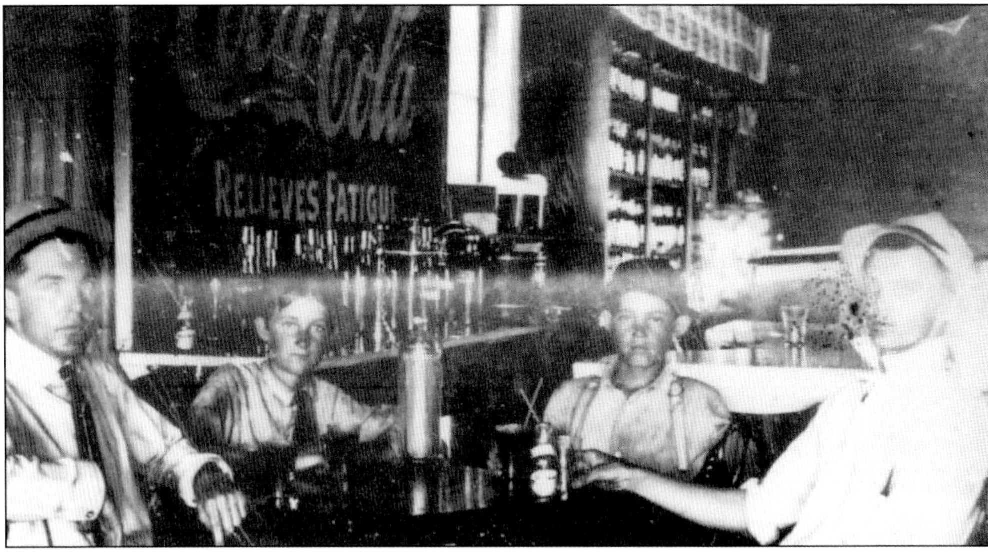

Friends gather for a cool drink in Corrie Hall's Canoe drugstore during the 1920s. The store, which also carried general merchandise, had a popular soda fountain that offered ice cream, sodas, Coca-Colas for 5¢, and coffee. On Saturdays, cows and hogs were butchered out back, making that day of the week the time when customers purchased their meat. (Courtesy of Carolyn Conn and Kevin McKinnley.)

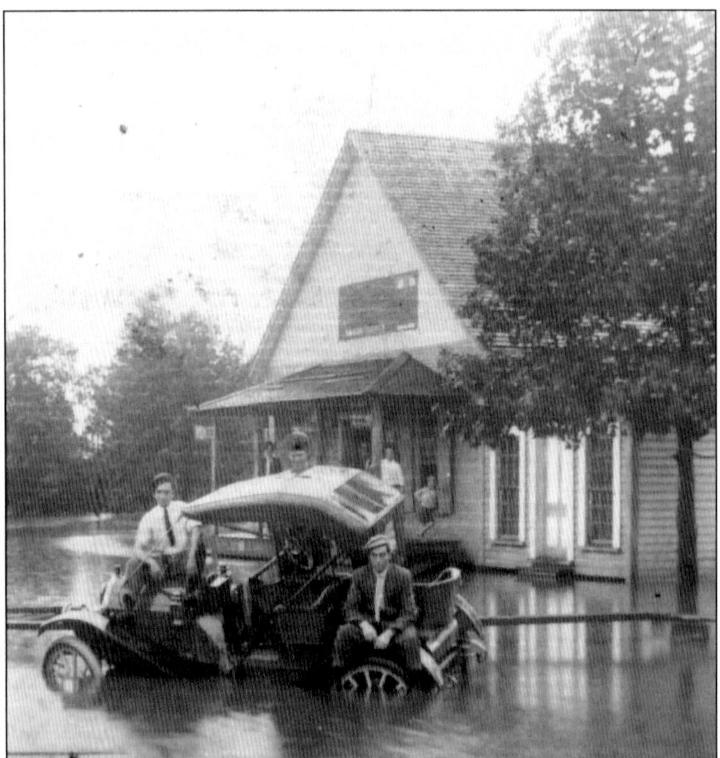

The flood of 1927 hit the area hard and left floodwaters up to the front door of Lowery's General Store in Canoe. The store sat next to the railroad and also housed the local post office. Canoe had flash floods because it was located on high, level grounds. (Courtesy of Carolyn Conn.)

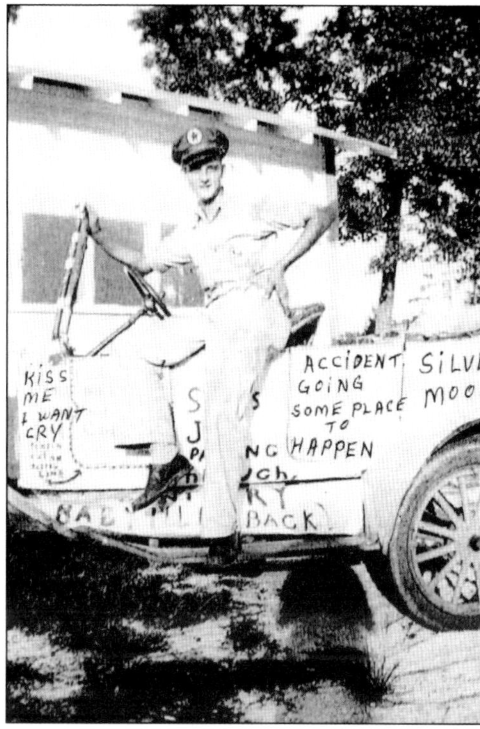

Mason Barnes, sitting atop his jalopy, is hanging out north of the twin bridges on Highway 31 near Pollard. By the messages, it seems that he wanted to make sure everyone saw him and that they knew he was single. (Courtesy of Jerry Simmons.)

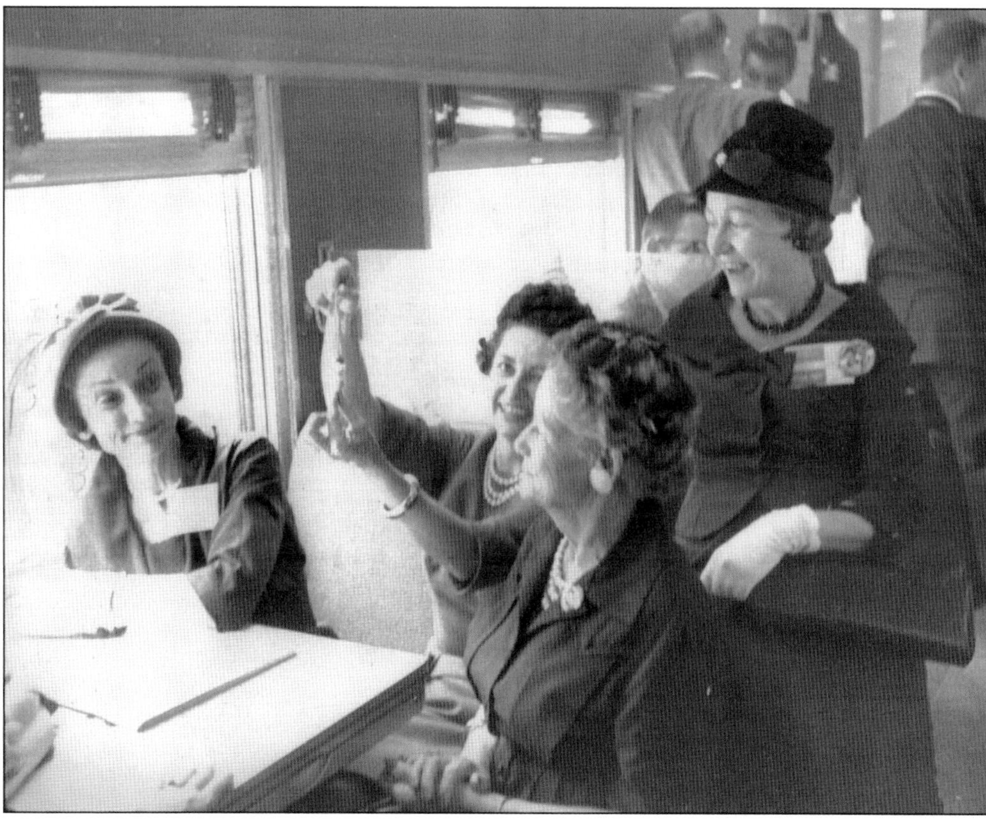

As part of the Kennedy/Johnson campaign in 1960, Lady Byrd Johnson stops in Flomaton with Lyndon B. Johnson, who was the candidate for vice president that year. Teresa Weaver, the wife of Flomaton mayor L.D. Warren, is also pictured. (Courtesy of Jerry Fischer.)

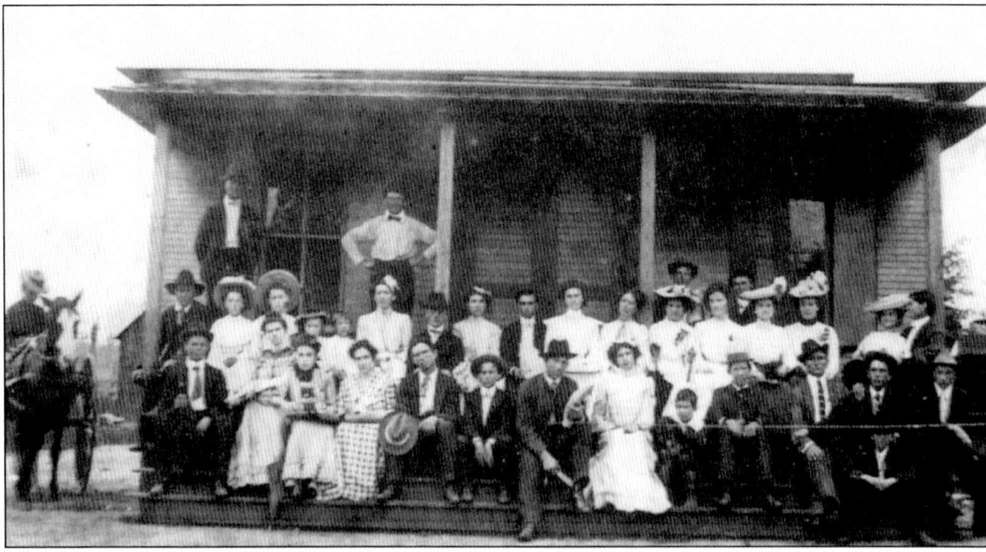

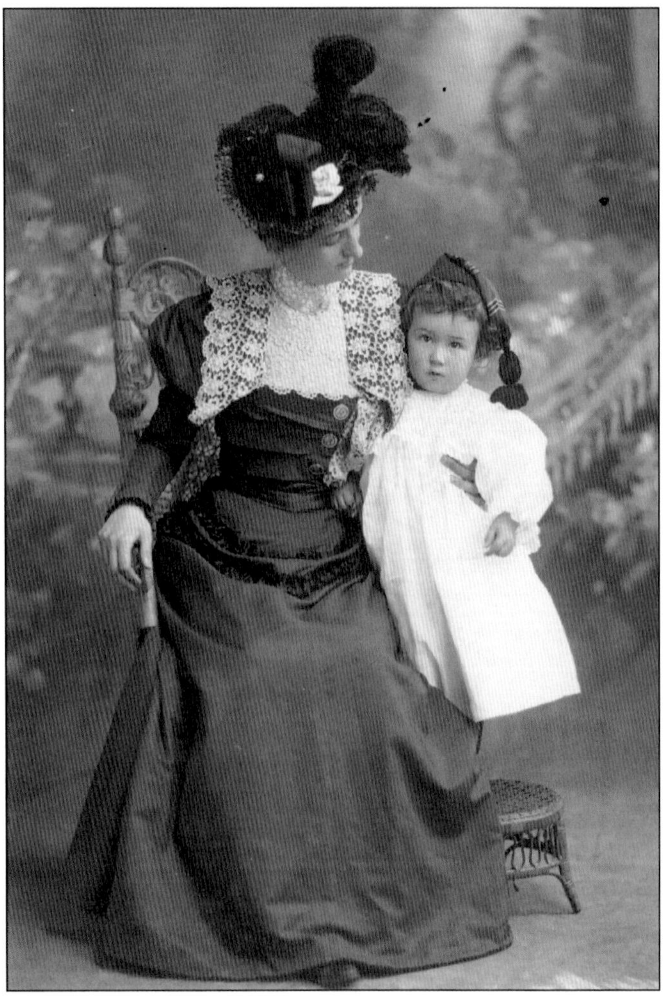

On Saturday nights in 1918 or 1919 in Canoe, the town's young people often gathered to dance in front yards to music provided by a radio. The drink of choice at the time was Coca-Cola. (Courtesy of Carolyn Conn and Kevin McKinnley.)

Harriet Carney Patterson and her son (William) Marshall Patterson are pictured in 1890. His father was Harry Howard "H.H." Patterson, who was a businessman and community leader in Atmore. H.H. helped to organize the original Atmore Chamber of Commerce. (Courtesy of Marshall Rogers.)

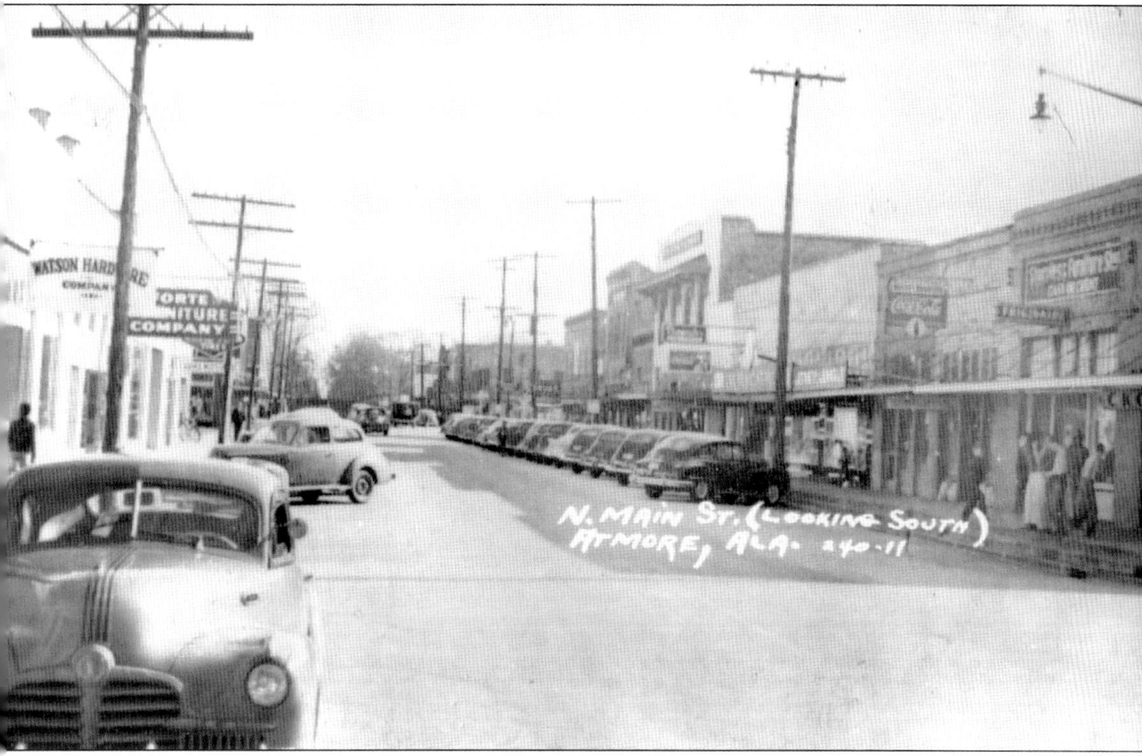

Another postcard of Atmore from the early 1940s is looking south on North Main Street. Doctors McKinley and Webb occupied the second floor above the furniture and hardware stores on the left side of the street. The town drugstore was run by Billy Holley, who later became mayor. (Courtesy of Fritz Currie.)

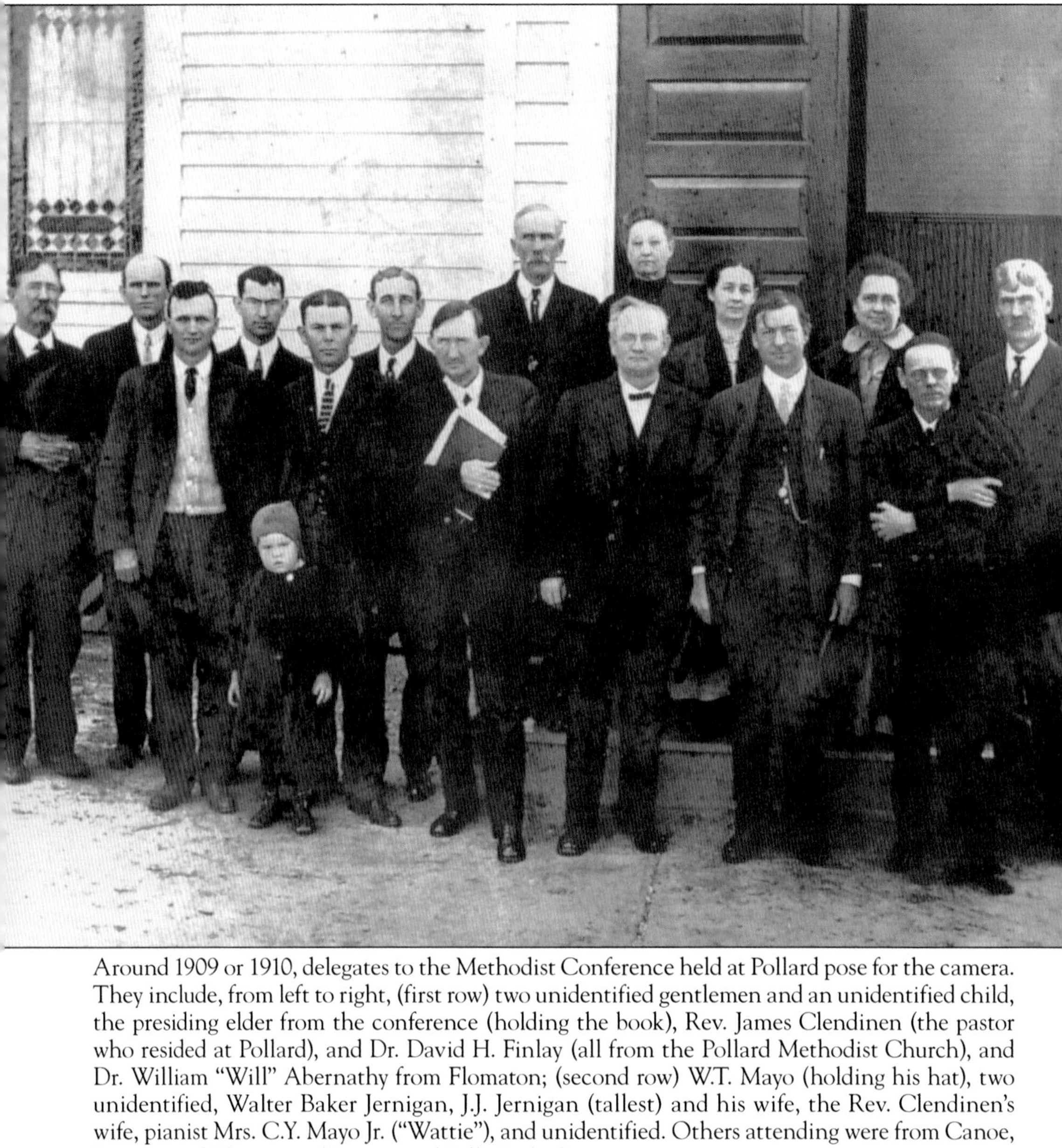

Around 1909 or 1910, delegates to the Methodist Conference held at Pollard pose for the camera. They include, from left to right, (first row) two unidentified gentlemen and an unidentified child, the presiding elder from the conference (holding the book), Rev. James Clendinen (the pastor who resided at Pollard), and Dr. David H. Finlay (all from the Pollard Methodist Church), and Dr. William "Will" Abernathy from Flomaton; (second row) W.T. Mayo (holding his hat), two unidentified, Walter Baker Jernigan, J.J. Jernigan (tallest) and his wife, the Rev. Clendinen's wife, pianist Mrs. C.Y. Mayo Jr. ("Wattie"), and unidentified. Others attending were from Canoe, Fannie, and Flomaton. (Courtesy of Sally Finlay.)

W.A. "Billy" Finlay, son of Pollard mayor H.D. Finlay, is standing outside the Pollard Station depot in the 1920s. His father owned the Pollard Gin & Produce Company that purchased produce from local farmers and shipped to wholesale grocery companies as far away as Cincinnati, Ohio, Louisville, Kentucky, and Nashville, Tennessee, as well as to Mobile, Alabama, and Pensacola, Florida. All of the produce shipped on the Louisville & Nashville (L&N) Railroad that ran through the town. (Courtesy of Sally Finlay.)

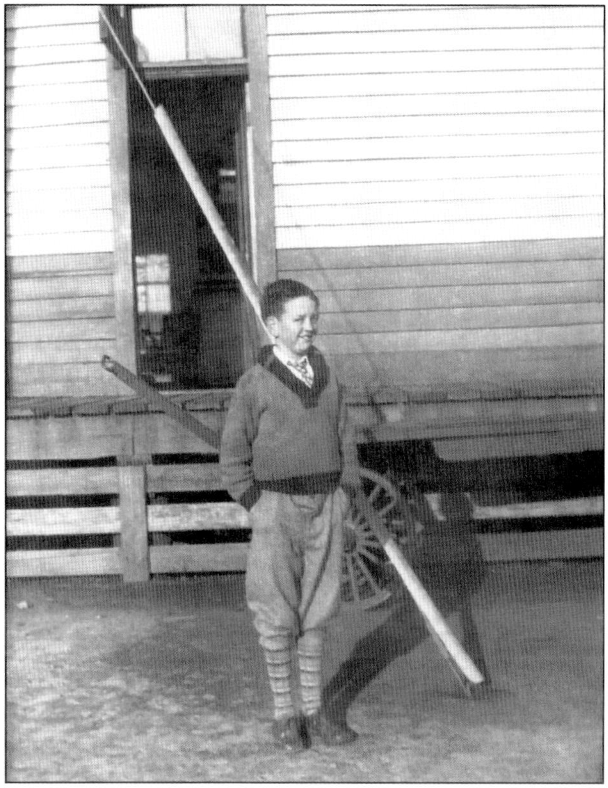

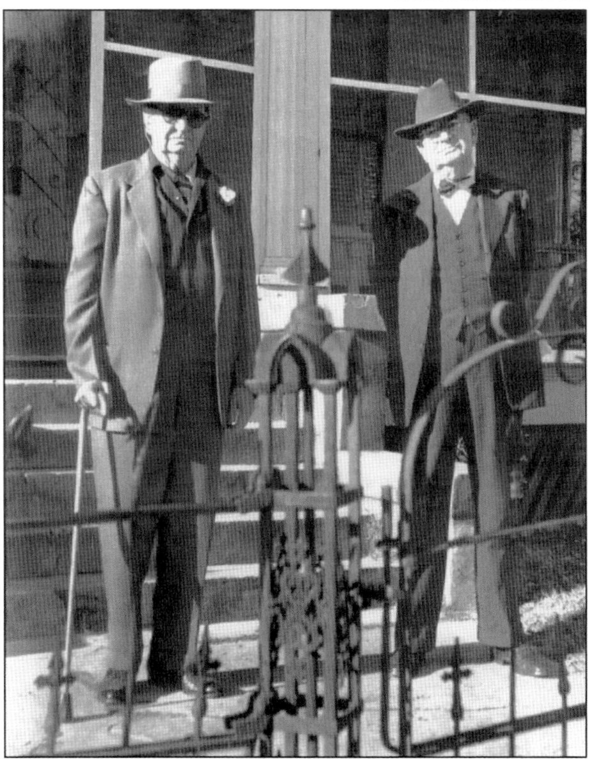

Mayor of Pollard H.D. Finlay Sr. (right) and his brother Dr. David Hume Finlay are outside of the H.D. Finlay home in Pollard. Dr. Finlay moved from Pollard to Blountstown, Florida, where he served as the doctor for the W.T. Neal Land and Timber Company. (Courtesy of Sally Finlay.)

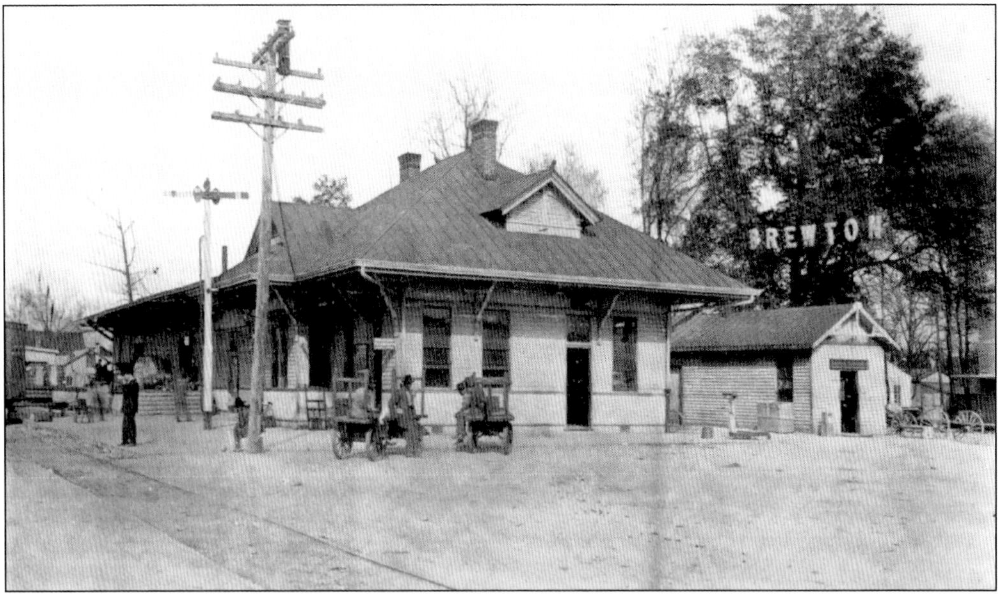

The L&N depot in Brewton was in the center of town. The town's letters hang in the trees in the background. The first railroad agent was Edmund Brewton. The area had been known as Crossroads and Newport, but once the railroad was built the town became known as Brewton. (Courtesy of Jerry Fischer.)

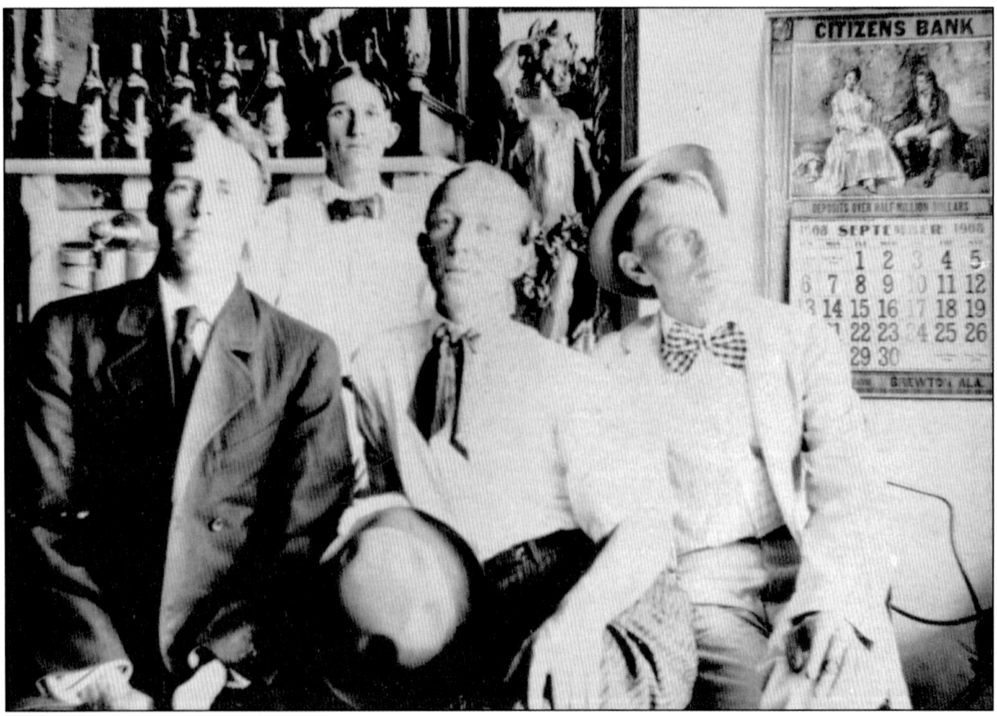

The date is September 1908, as evidenced on the calendar. This picture of a good-looking group of men was taken at the Palace Pharmacy in downtown Brewton. Burke Hayes is on the left, while Oscar Colley is in the middle, and Joe Booth, the soda fountain jerk, is in the back. The gentleman with the bow tie is unidentified. (Courtesy of D.W. McMillan Trust.)

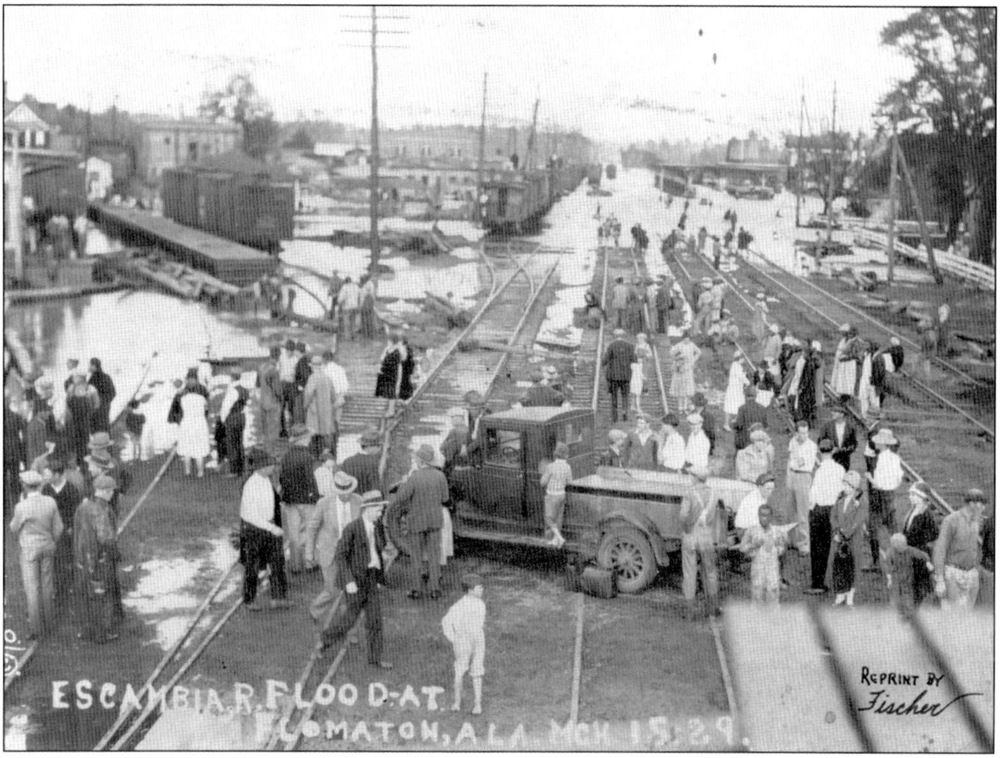

The biggest flood in Escambia County's history occurred in March 1929. All areas near the Conecuh River or the creeks and streams that flow into the river were flooded. Flomaton is near Big Escambia Creek. This picture is looking east along the tracks toward downtown Flomaton and Big Escambia Creek. (Courtesy of Jerry Fischer.)

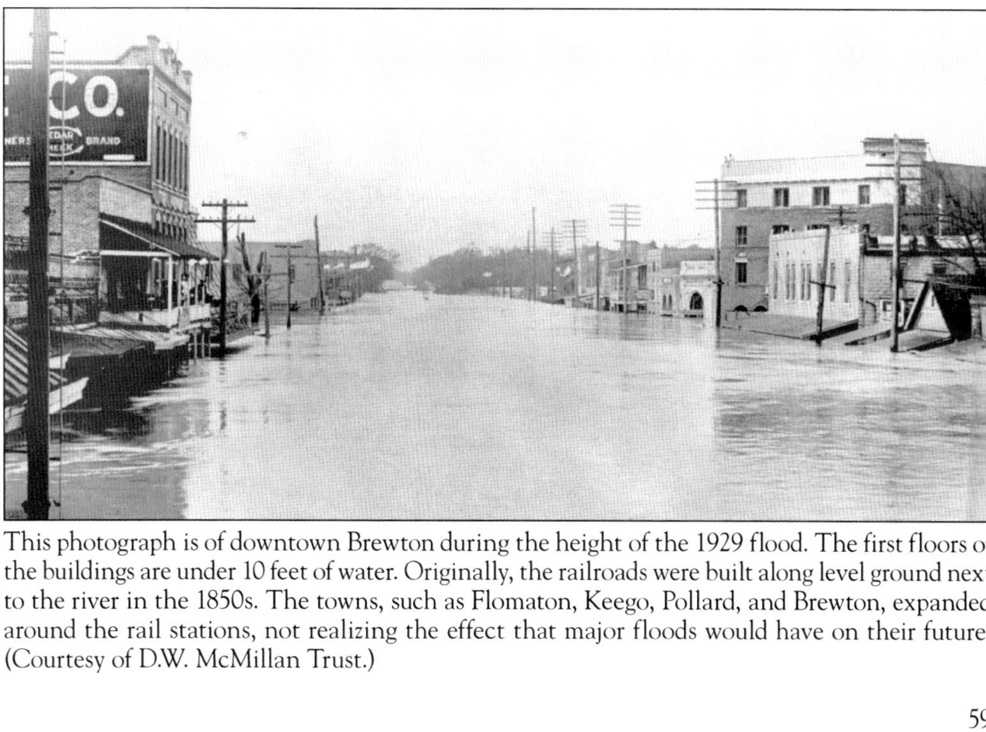

This photograph is of downtown Brewton during the height of the 1929 flood. The first floors of the buildings are under 10 feet of water. Originally, the railroads were built along level ground next to the river in the 1850s. The towns, such as Flomaton, Keego, Pollard, and Brewton, expanded around the rail stations, not realizing the effect that major floods would have on their future. (Courtesy of D.W. McMillan Trust.)

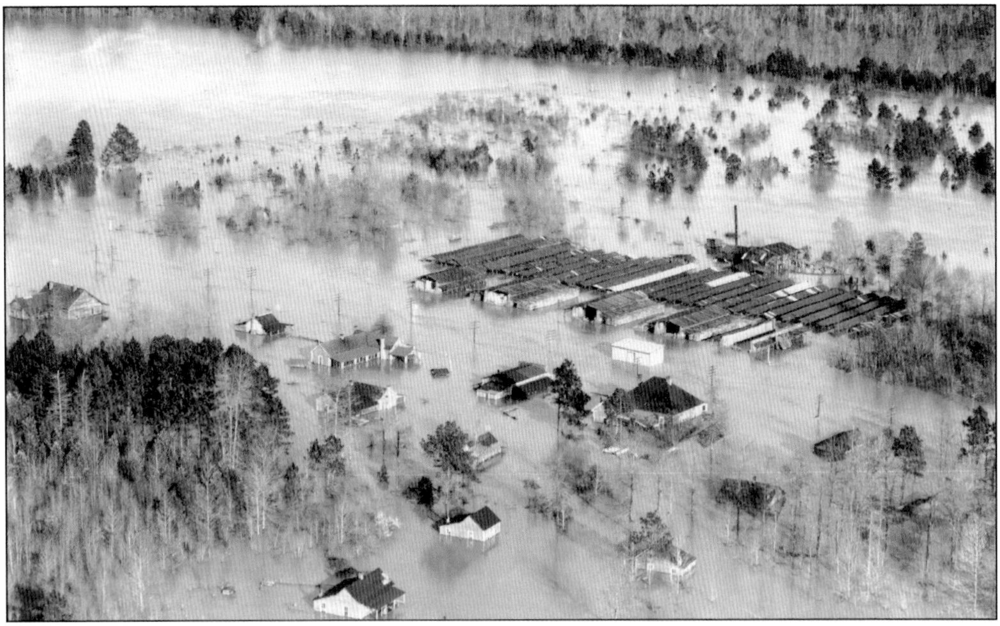

This picture shows Keego during the height of the 1929 flood. The large group of buildings is the Keego Brick Company, which was started in 1899 by E.M. Lovelace and W.H. Strong. Ben L. "Fee" Martin later became owner of the business and made many bricks from local clay until recent time. Keego was located on the L&N Railroad. (Courtesy of Montgomery Alabama Department of Archives and History.)

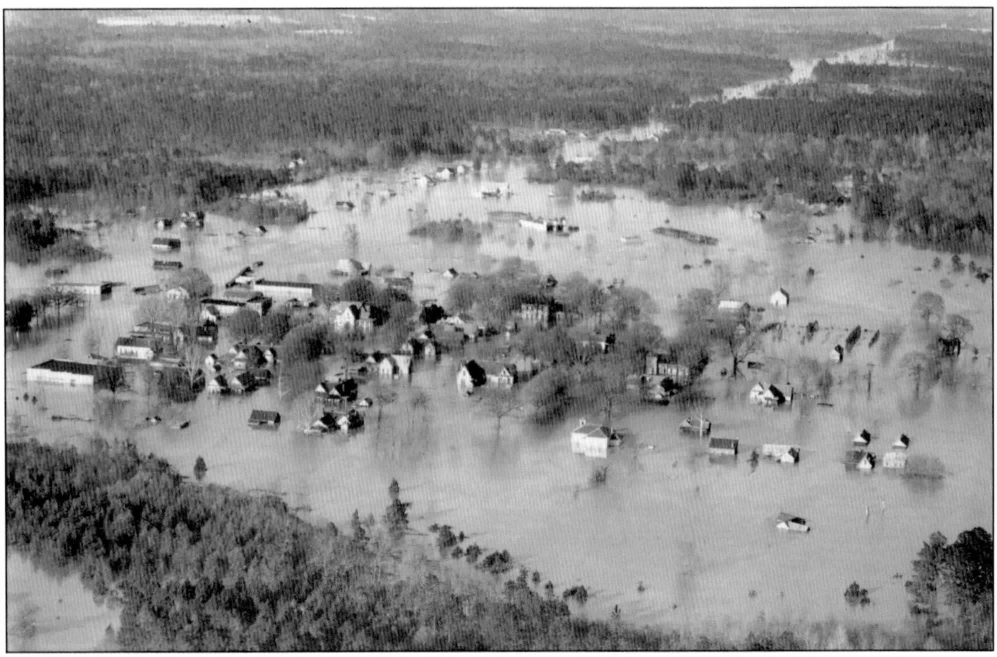

This aerial shot of Pollard during the 1929 flood shows Pollard at its peak. The 1929 flood had long-lasting effects on all Escambia County towns except for Atmore. The Great Depression followed soon after, and Pollard never recovered. (Courtesy of Montgomery Alabama Department of Archives and History.)

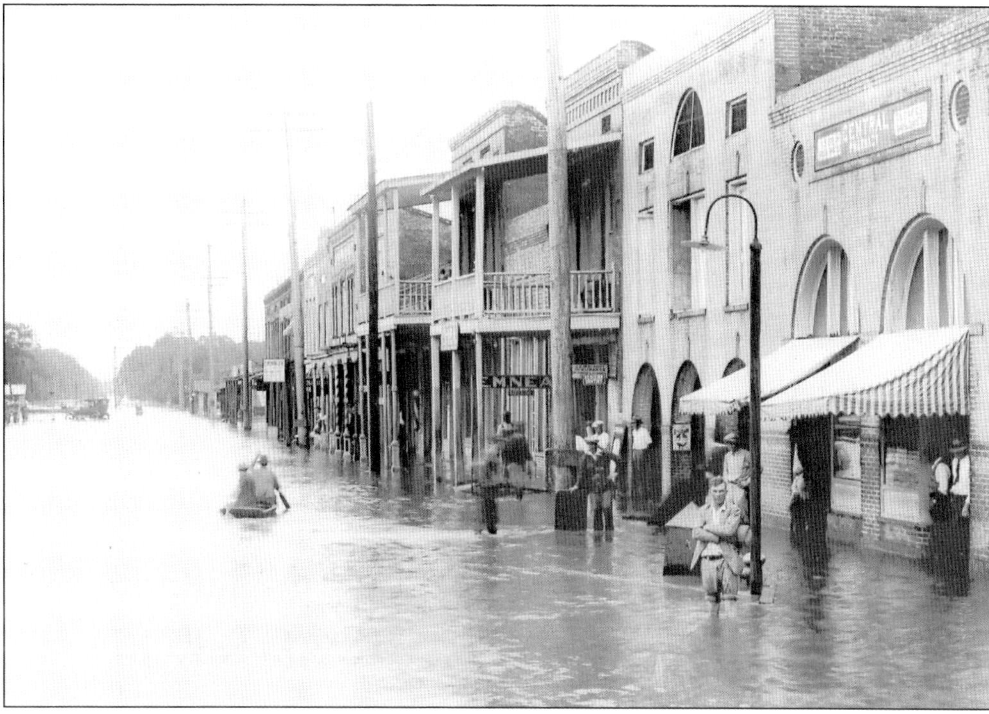

Pictured is the 1929 flood of downtown Brewton looking to the south. The waters had started to recede at this point as they flowed to the south. (Courtesy of D.W. McMillan Trust.)

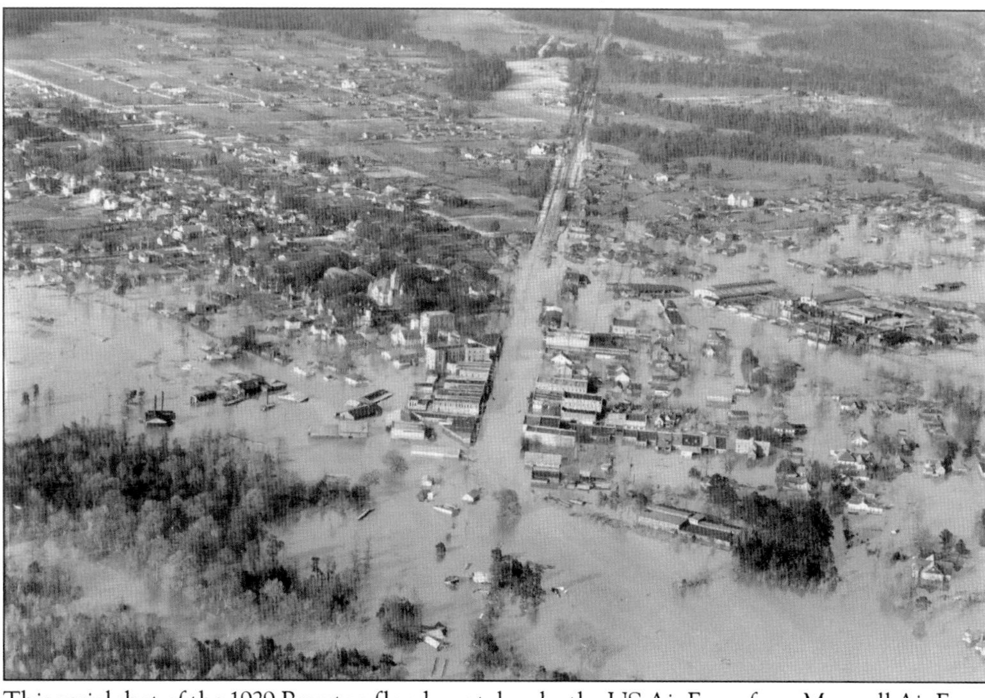

This aerial shot of the 1929 Brewton flood was taken by the US Air Force from Maxwell Air Force Base in Montgomery, Alabama. T.R. Miller Mill Company is to the top right, and the sawmill is to the left. (Courtesy of Montgomery Alabama Department of Archives and History.)

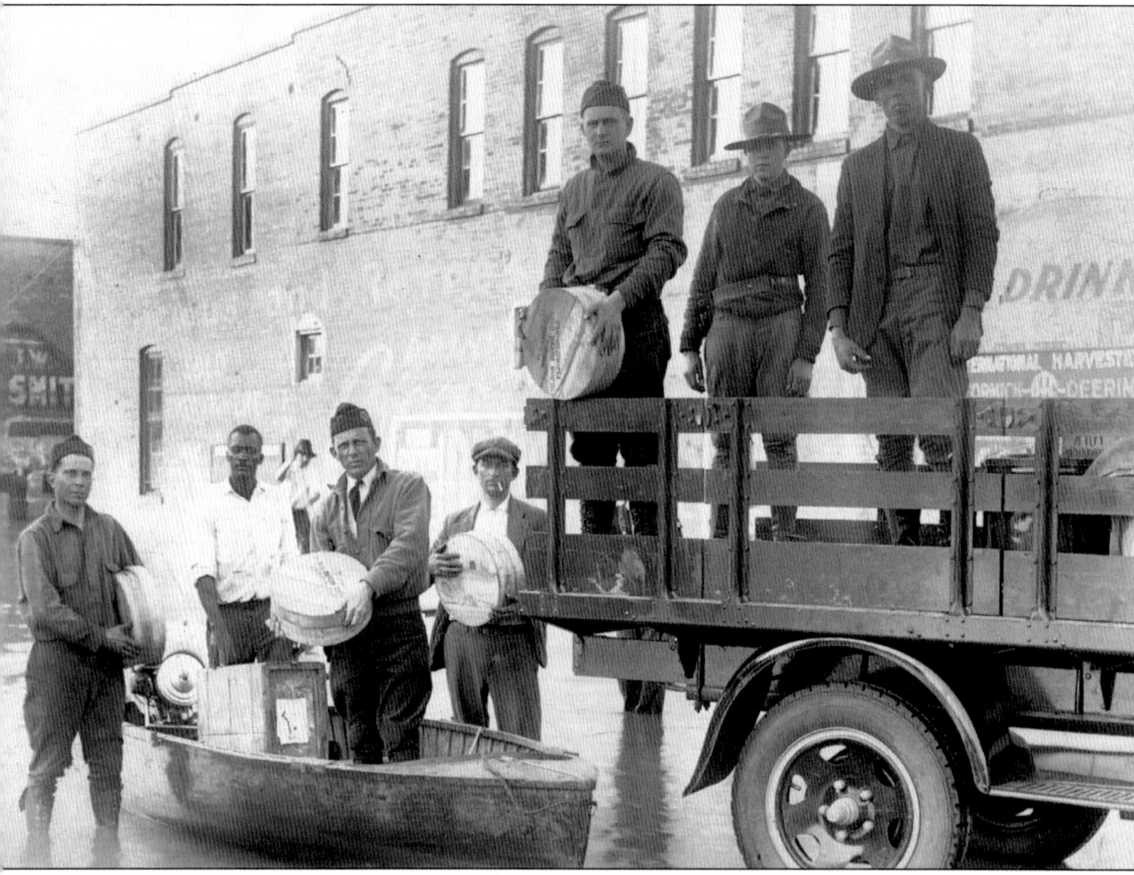

During the 1929 flood, the military dropped bread from airplanes to the people of Brewton, Pollard, and Flomaton. Other troops delivered food items by truck and then by boat, as shown here in Flomaton. Included in the image are C.C. Fountain, Terrell Weaver, and Edgar Downing. (Courtesy of D.W. McMillan Trust.)

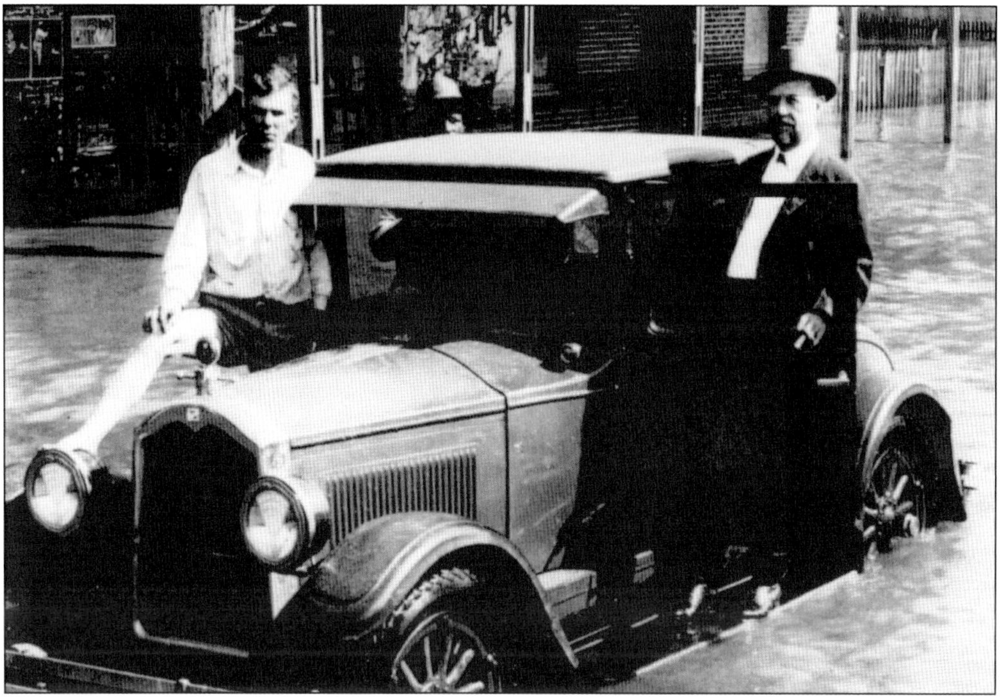

Dr. Russell A. Smith (right) is posing for a picture during the 1929 flood in Brewton. Dr. Smith was the cofounder of the Escambia County Hospital and the mayor of Brewton for a period. (Courtesy of D.W. McMillan Trust.)

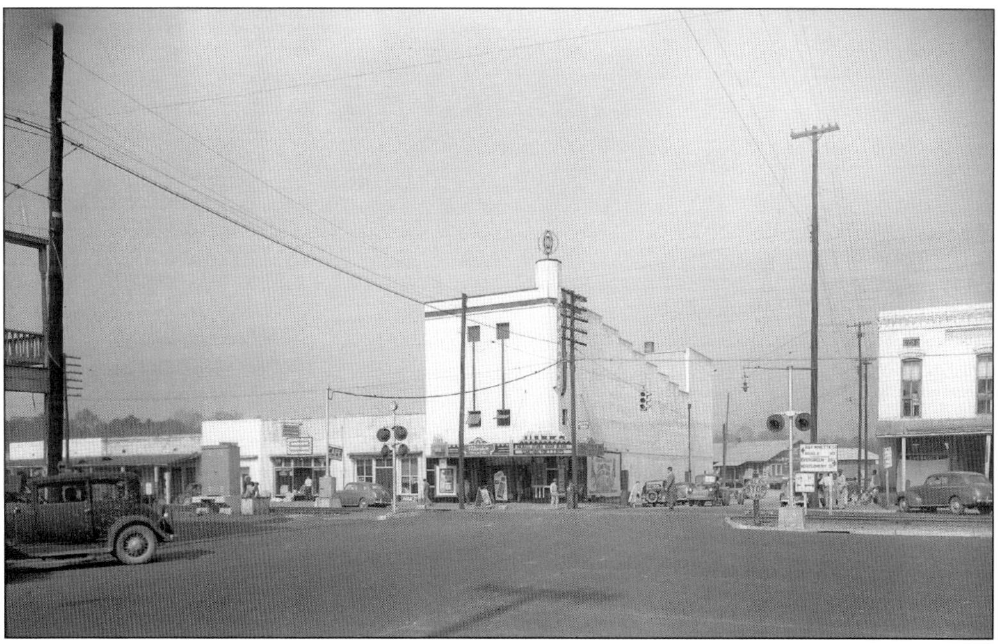

This photograph is of downtown Brewton at the intersection of US Highways 31 and 29 during dryer times in 1944. The tall building is the Ritz Theater. The movie showing is *Something for the Boys*, starring Carmen Miranda, Michael O'Shea, and Vivian Blaine. (Courtesy of D.W. McMillan Trust.)

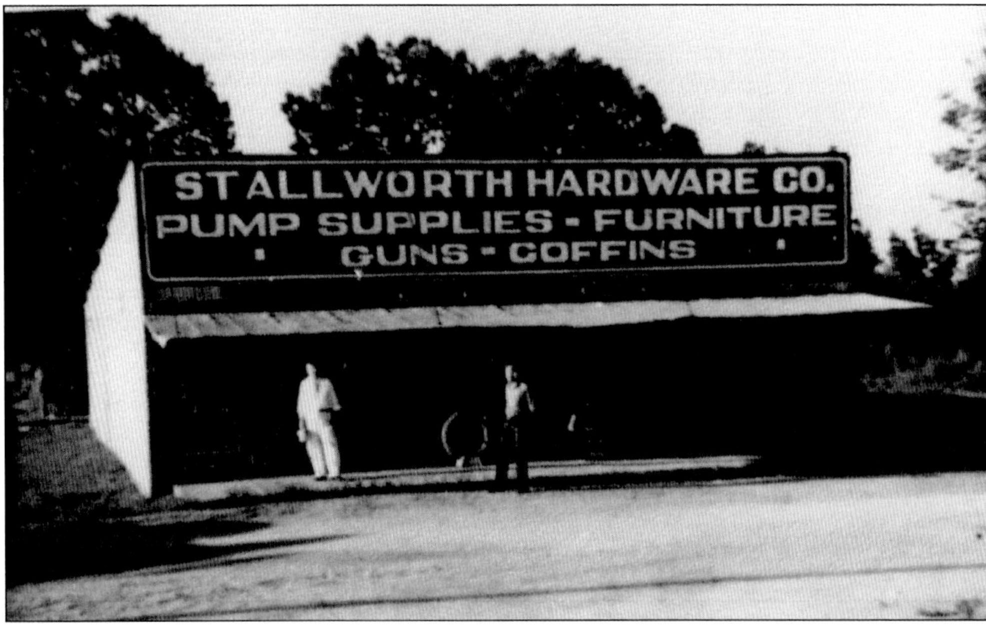

Perhaps the most memorable sign in the area belonged to the Stallworth Hardware Company in Canoe, which offered the unusual combination of guns and coffins. Coffins were routinely kept on display, as customers would choose one after a death in the family and then have it sent to the local undertaker. Out-of-towners often knew Canoe as "that place with the gun and coffin store." The back of the building was the business home of "Bradley the Blacksmith," who often rewarded well-behaved children by letting them turn on the air blower in his shop. (Courtesy of Carolyn Conn.)

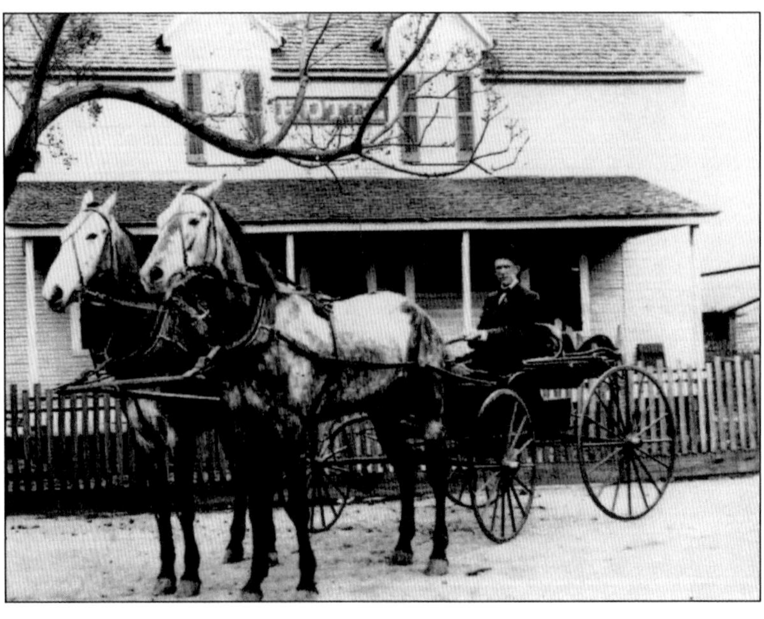

The Knowles Hotel in Canoe was a short distance from a mineral spring, which was a big attraction. Hotel patrons frequented the spring (now known as Canoe Creek), while trains passing through the area often stopped to fill up on mineral water. (Courtesy of Carolyn Conn and Kevin McKinnley.)

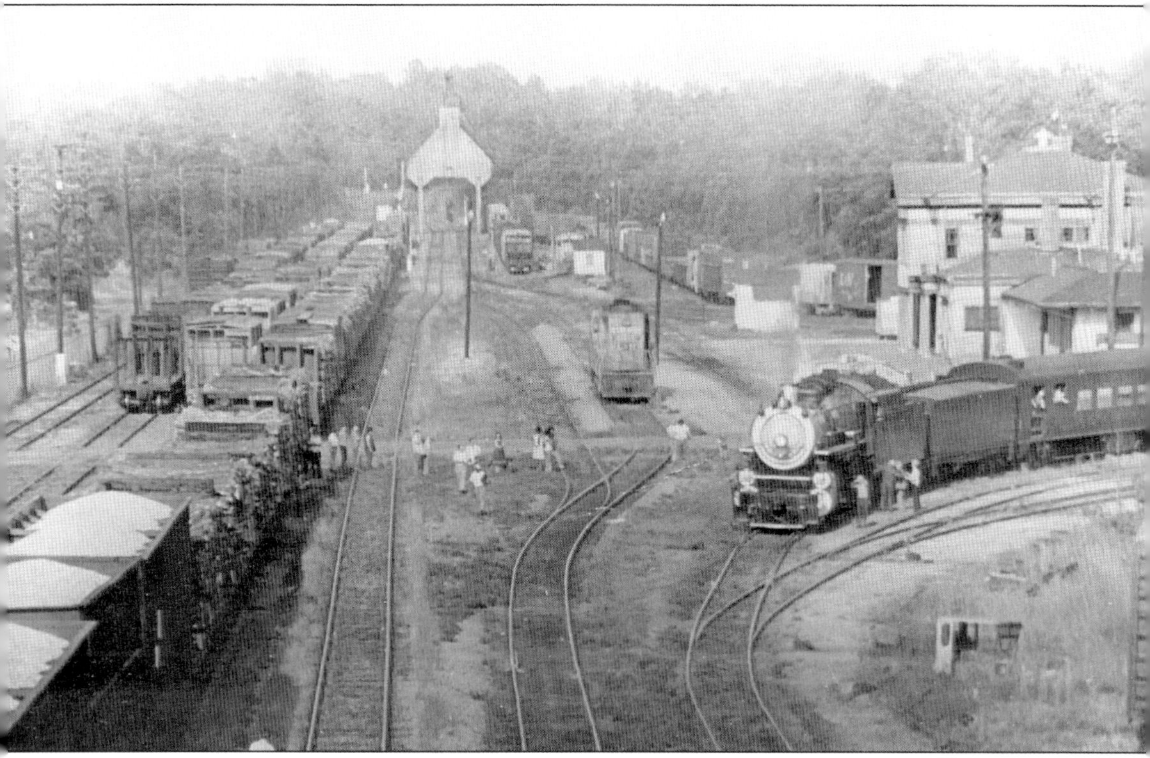

Flomaton became the junction of the Mobile & Great Northern Railroad with the Alabama & Florida Railway after the Civil War. These lines were later to become the Louisville & Nashville Railroad. (Courtesy of Lydia Grimes.)

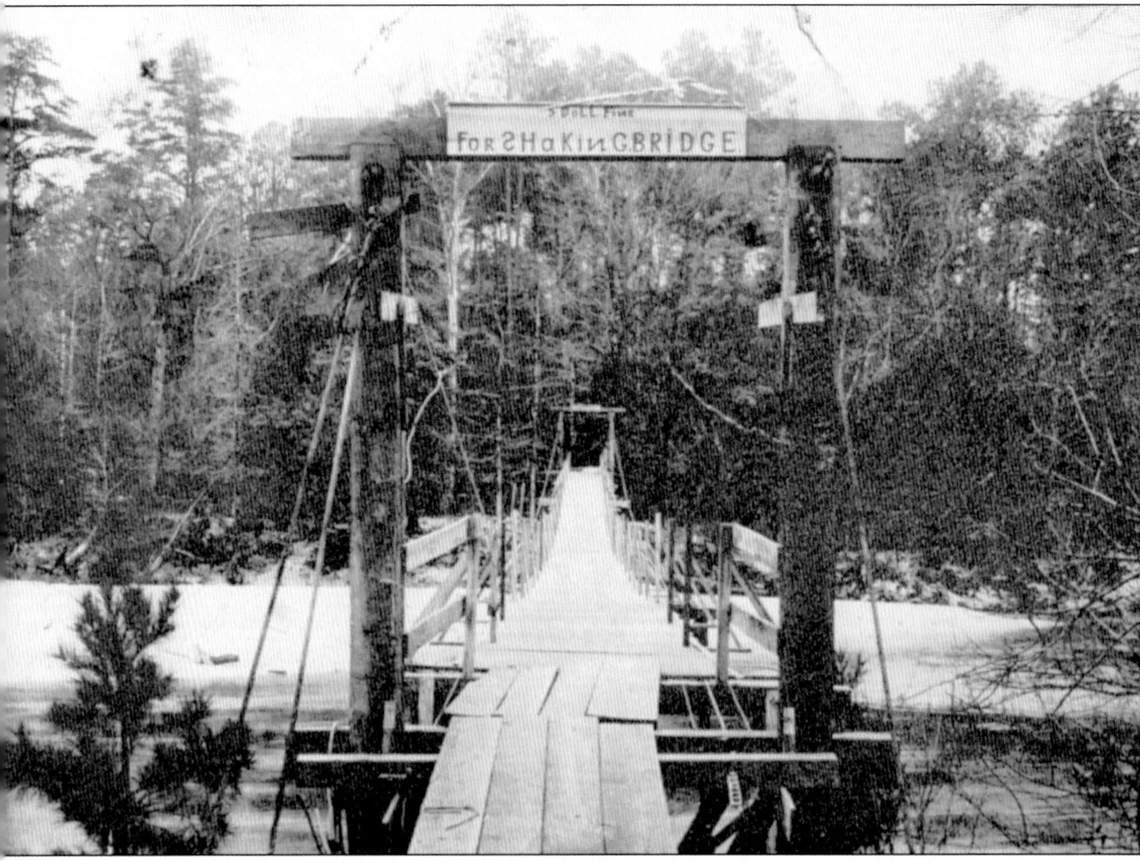

This sign reads, "Five doll. fine for shaking bridge." This is the suspended bridge going across Murder Creek between Brewton and East Brewton. It is difficult to be sure if anyone ever had to pay a fine. (Courtesy of D.W. McMillan Trust.)

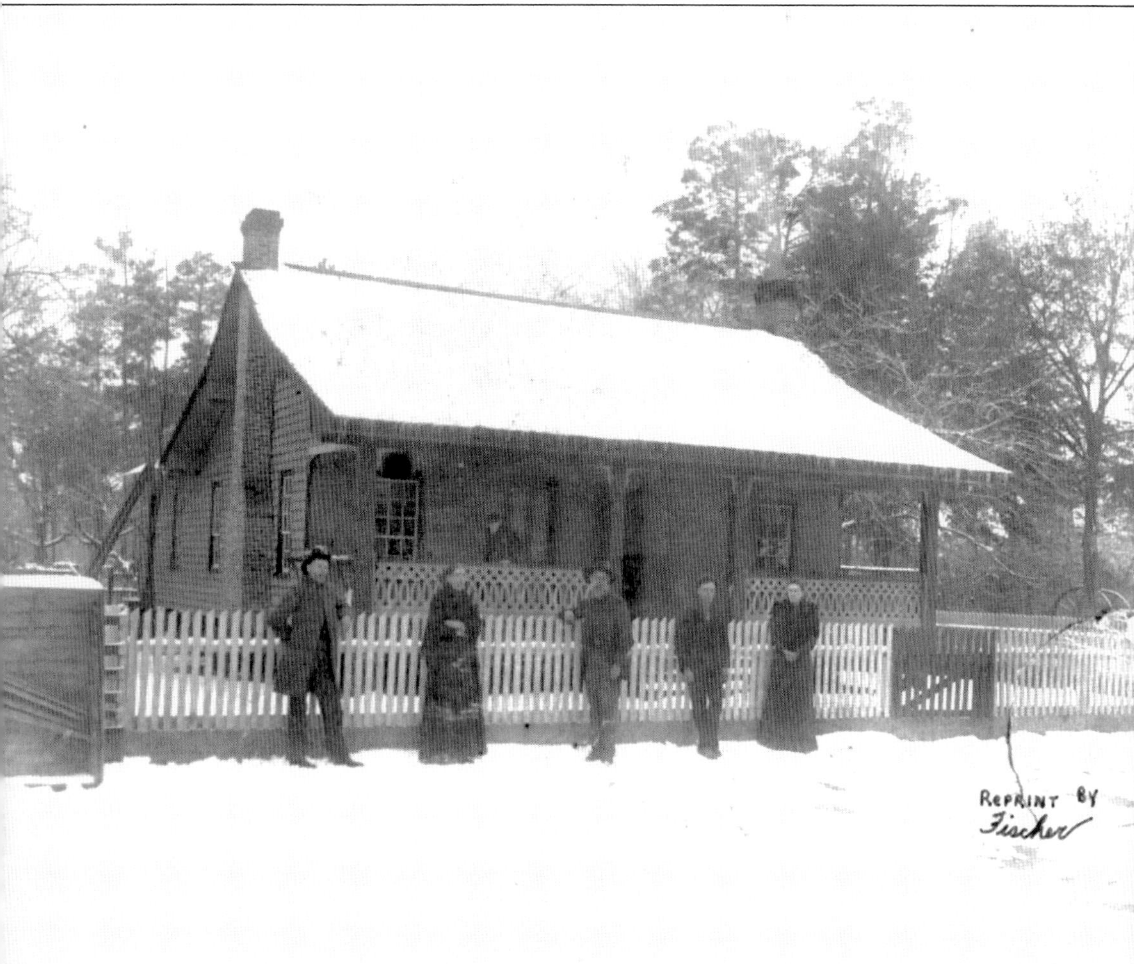

South Alabama will occasionally receive a real snowstorm, and this shot of the Herrington-Jernigan home at Herrington Springs is proof. Herrington was a bustling town on the L&N Railroad line, but it now consists of only a few homes. (Courtesy of Jerry Fischer.)

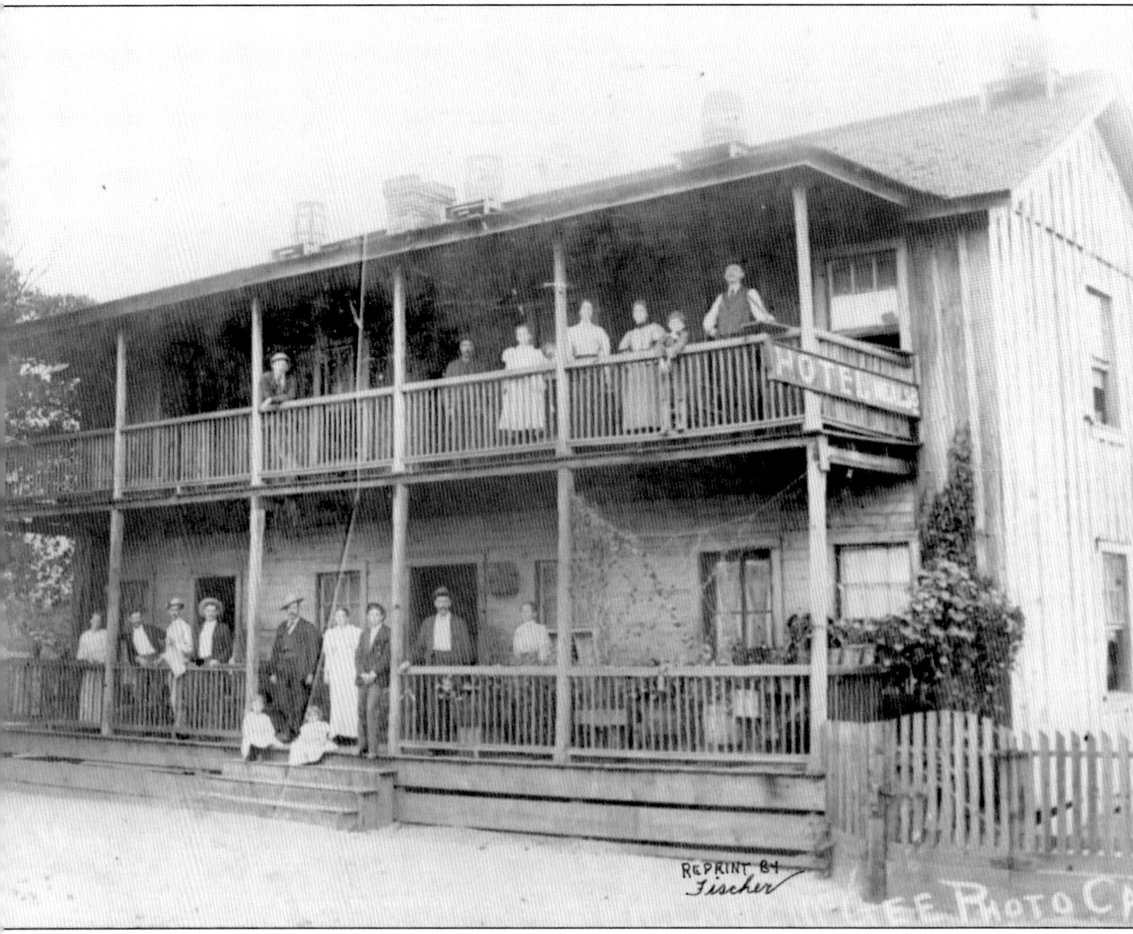

The water drums on top of the hotel at the depot in Flomaton are for fire protection. They are filled by rainwater, and a rope is used to overturn the drums in the case of a fire. (Courtesy of Jerry Fischer.)

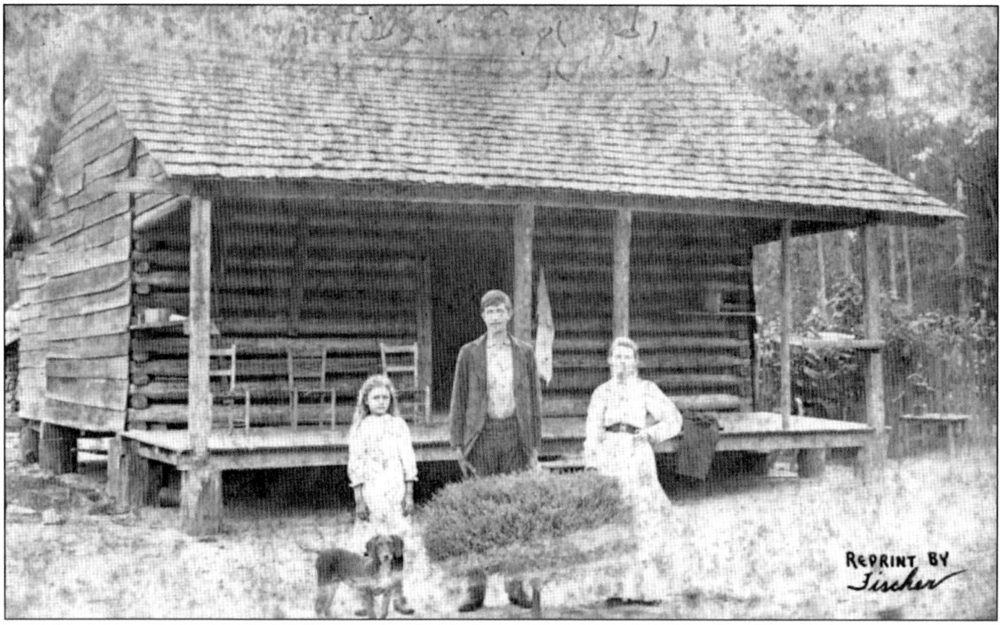

This log cabin in Pollard is a good example of the early days of hard living in south Alabama. There are no windows in the cabin, which sits on blocks of heart pine to keep out the termites. Pictured are John Lindsey, Mattie Lindsey (left), and Mae Bell Lindsey. (Courtesy of Jerry Fischer.)

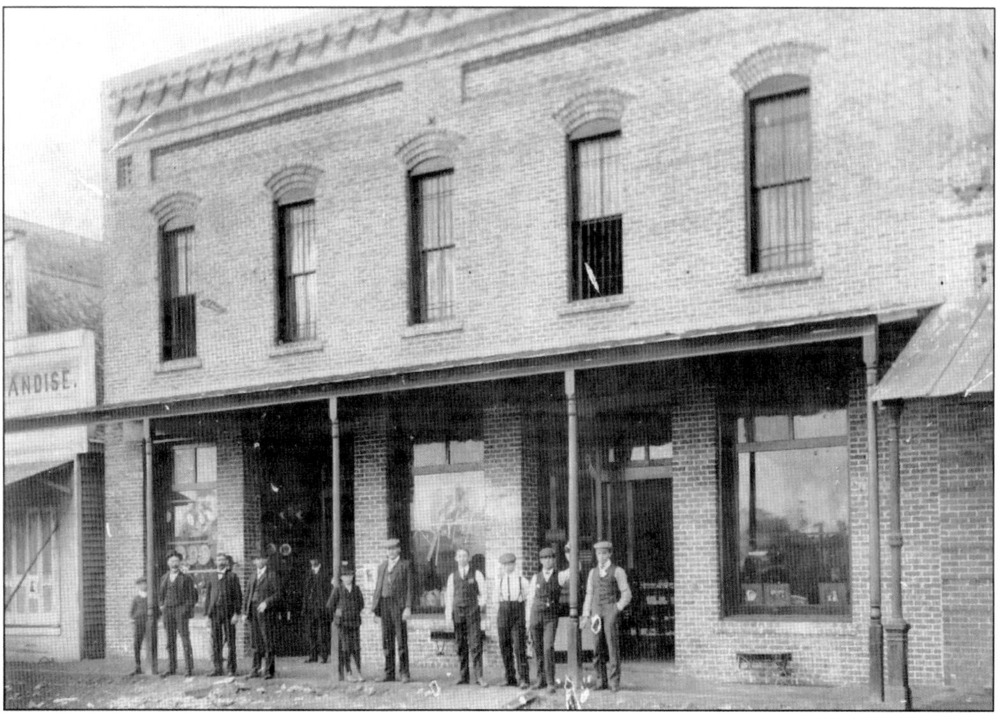

In 1905, People's Mercantile Company was a major source of goods in Atmore. Legend has it that a local banker took money from the bank at which the store's funds were kept and lost it all gambling. The financial devastation resulted in People's Mercantile eventually becoming Southern Escambia Hardware. (Courtesy of Fritz Currie.)

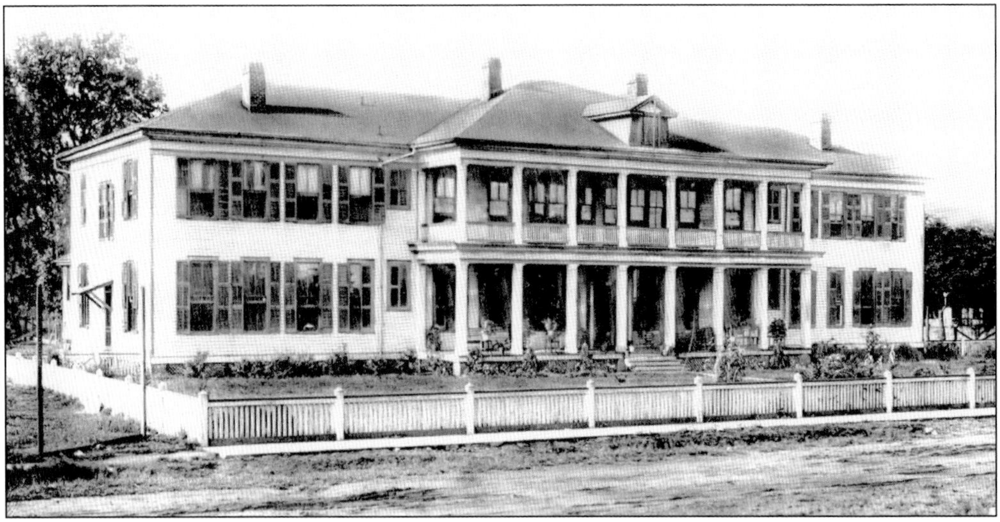

The Carney Hotel was a landmark in Atmore for many years. It was originally built by W.M. Carney and H. Carney in 1908 for $15,000. It then became the Crook Hotel in 1913 and finally the Norman Hotel before it was demolished. The hotel was facing the depot looking north about where the United Bank of Atmore now stands. (Courtesy of John Lewis.)

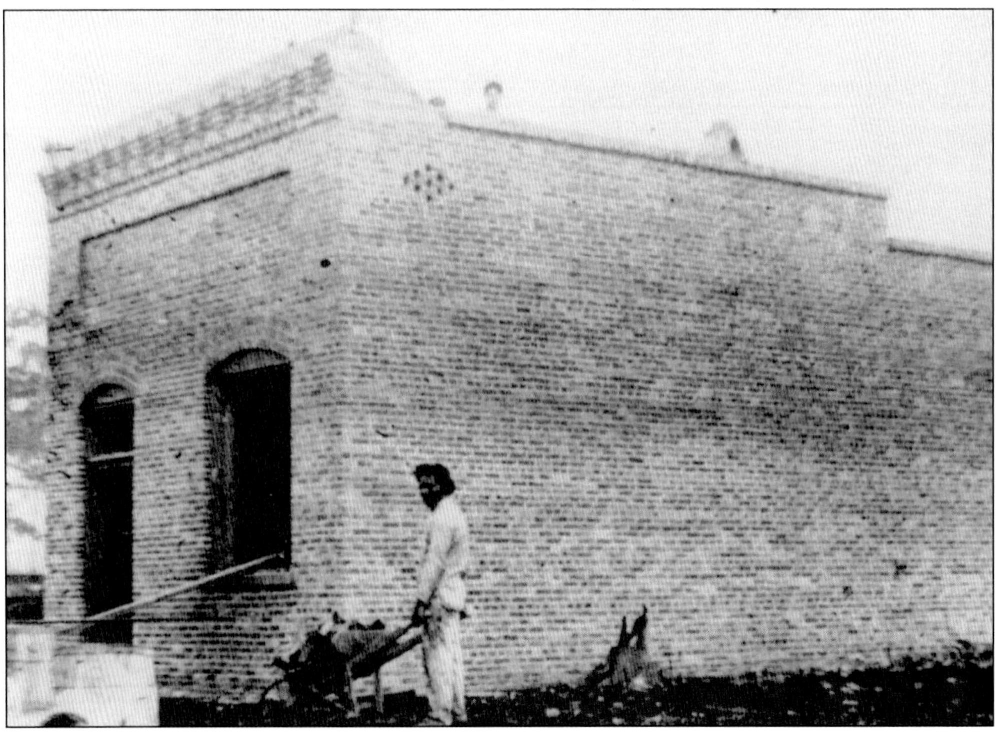

This is the late-1920s construction of the bank in Canoe that never opened, most likely because of the stock market crash of 1929 and the Great Depression. While the interior of the bank was never completed, the owner left a giant pot-bellied safe inside. The abandoned building became a favorite playhouse for neighborhood children and a natural setting for games of "cops and robbers" and "cowboys and Indians." The building later enjoyed new life as Nixon's Grocery Store and as an elementary school. (Courtesy of Carolyn Conn and Kevin McKinnley.)

Ed Leigh McMillan and Iva Lee Miller were married in 1911. They both loved the history of the area and are responsible for saving photographs and writing down all that they could for future generations. Ed Leigh enjoyed a long legal and business career, while Iva Lee was involved in civic and community affairs. (Courtesy of D.W. McMillan Trust.)

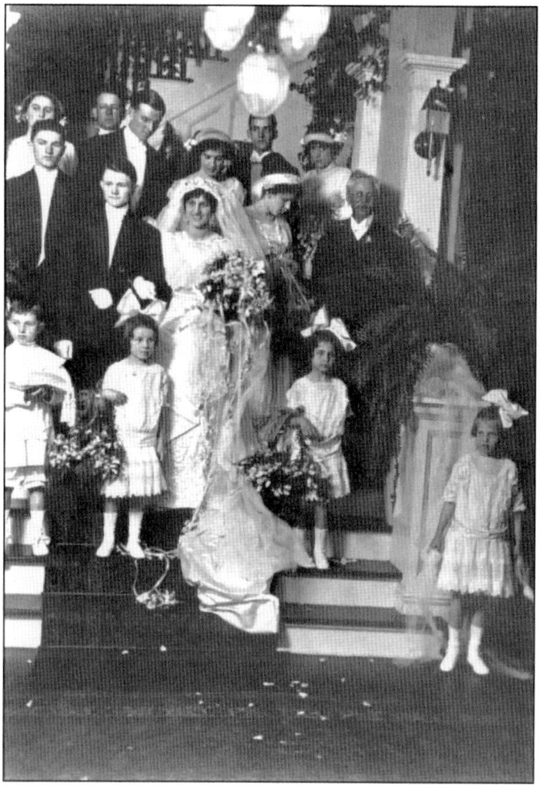

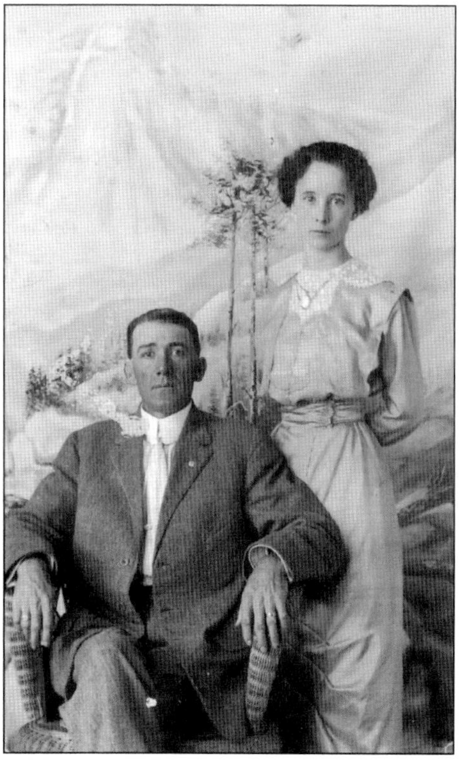

Wallace became a very busy town on the Pensacola & Selma Railroad, which later became part of the L&N Railroad. The Kelly family had a cotton gin and bank in Wallace. Charles Sowell built a sawmill and eventually had 20 miles of log ditches constructed for his operations. (Courtesy of Escambia County Historical Society.)

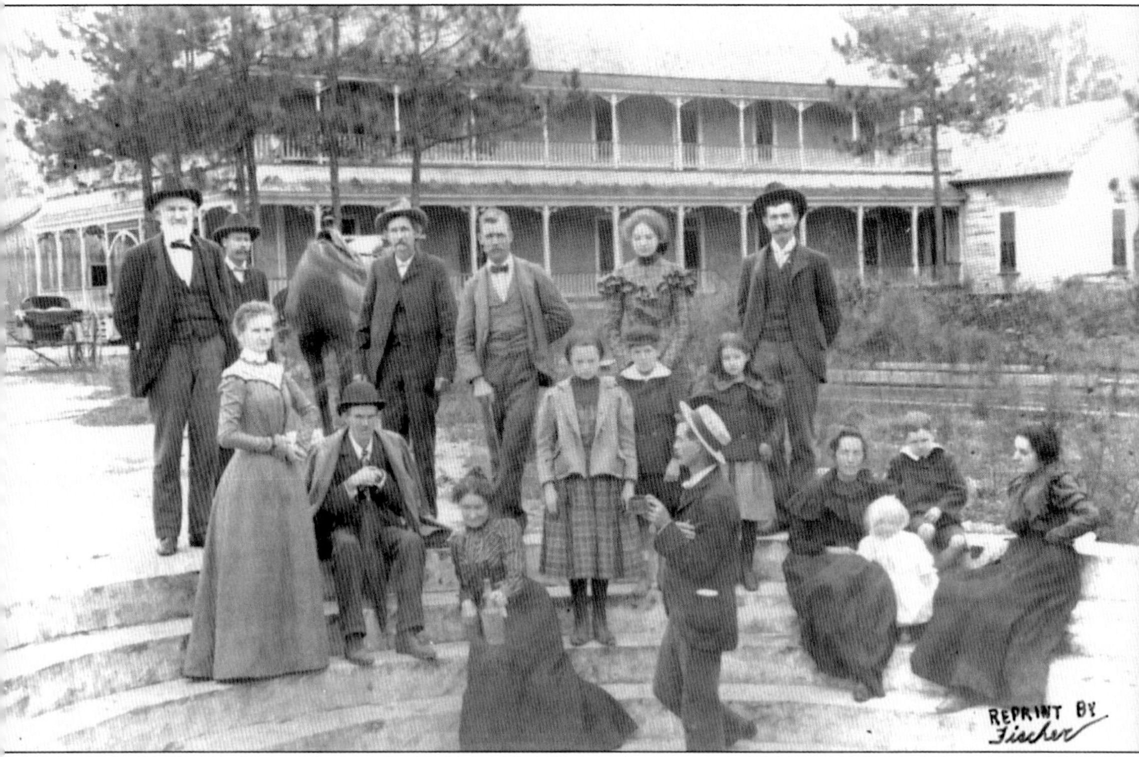

In the late 1800s, a mineral spring was found in Keego by James Herrington, who built a 22-room resort hotel named El Dorado. The water was full of iron and iron salts and was commonly known as "good for what ails you." Patrons could stay here in style for $4 a week room and board with access to the medicinal pool. It had a post office at one time. From left to right are (first row) Jessie Ghent and Charlie McGraw; (second row) Laura Lovelace, Byrus Butler, Lillie Bell Rankin, Morey Martin, Sumpie Martin, Jessie Lovelace Martin, "Fee" Martin, Billie Martin, and Bennie Lovelace; (third row) B.M. Lovelace, James Jernigan, Dr. J.E. Martin, unidentified, Navalou Martin, and Ernest Doors. (Courtesy of Jerry Fischer.)

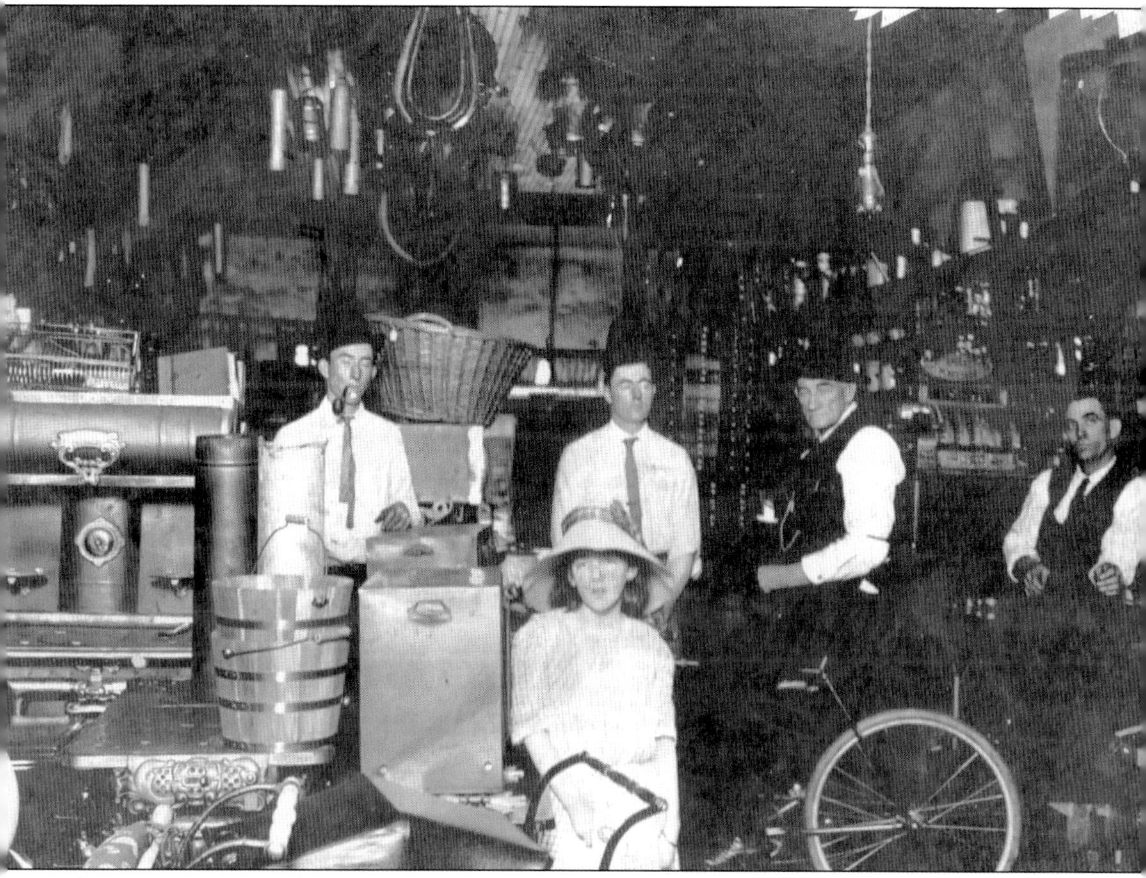

The Atmore Mercantile Company in downtown Atmore appears between 1911 and 1916. The general store sold anything and everything from the horse collars hanging from the ceiling to the baby carriages seen in the foreground. Pictured in back from left to right are Sam Byrne, Jay Marshall, Daniel Archibald Currie, and H.W. Currie; the young lady in the foreground is Dannie Currie. (Courtesy of Fritz Currie.)

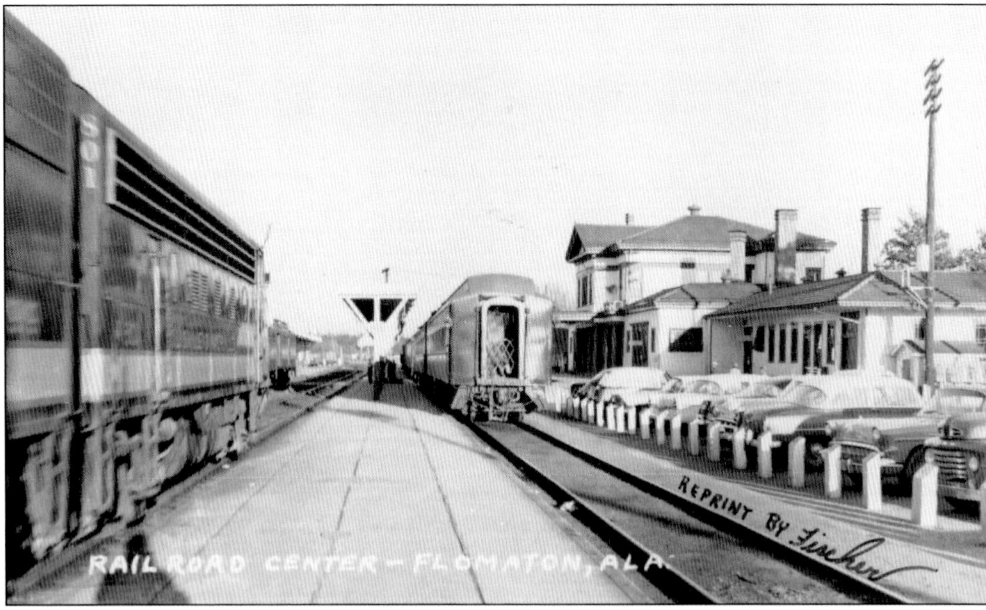

A passenger train is shown at the Flomaton Train Station. It was the central place for freight and passengers. The two previous stations burned down because it is believed that they were too close to the coal chute, so this structure was built a quarter mile away from the previous stations. (Courtesy of Jerry Fischer.)

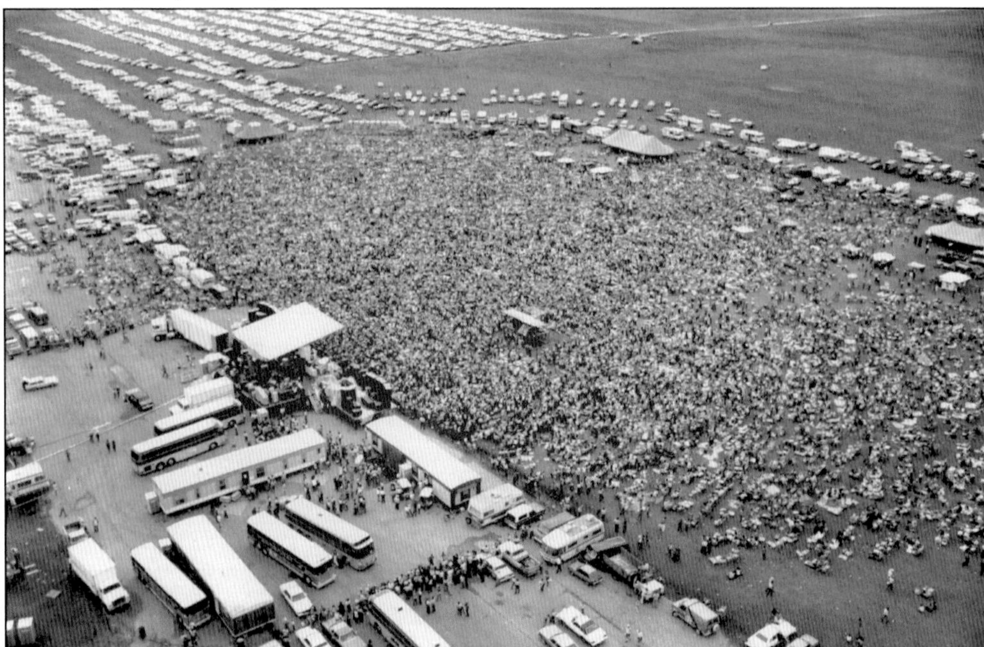

Local promoter and businessman Buddy Liles, along with Eli Ball of Nashville, put together an outdoor concert known as the Harvest Jam at the Brewton Airport. The concert was in October 1981, and over 35,000 people attended. Big-name artists included the Oak Ridge Boys (with Brewton native William Lee Golden), Charlie Daniels, Alabama, Delbert McClinton, and Sylvia. (Courtesy of Buddy Liles.)

Five

BUSINESS AND INDUSTRY

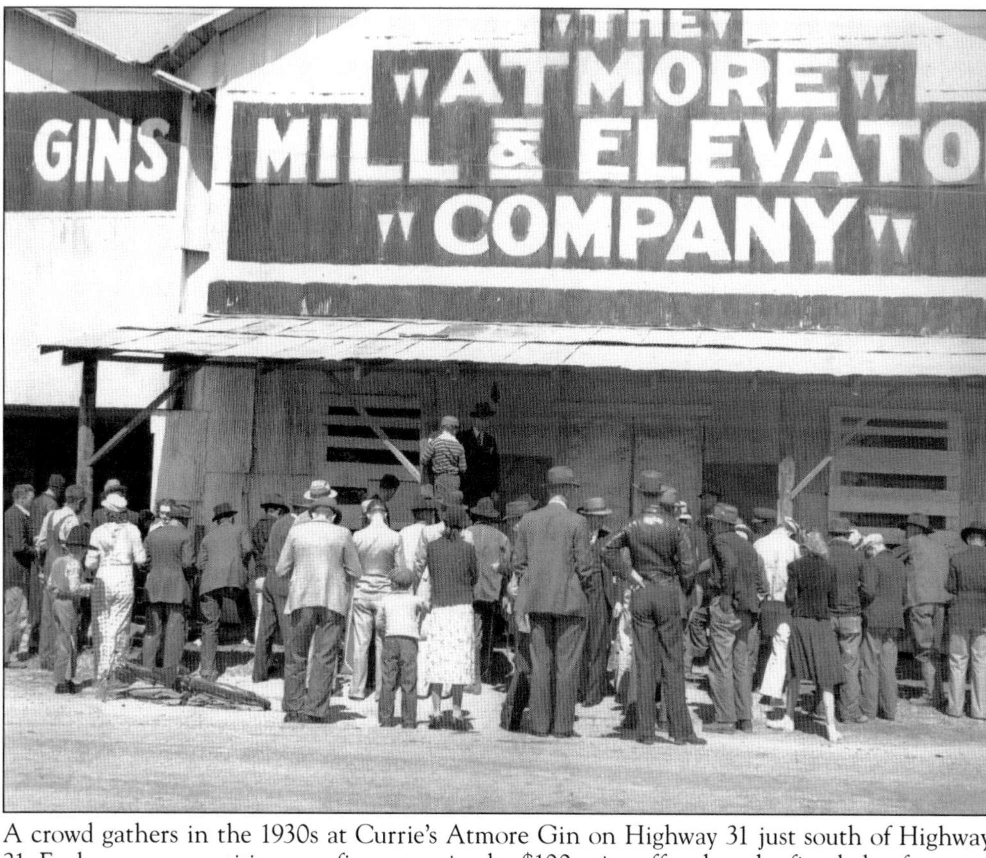

A crowd gathers in the 1930s at Currie's Atmore Gin on Highway 31 just south of Highway 21. Each year, competition was fierce to win the $100 prize offered to the first bale of cotton. (Courtesy of Fritz Currie.)

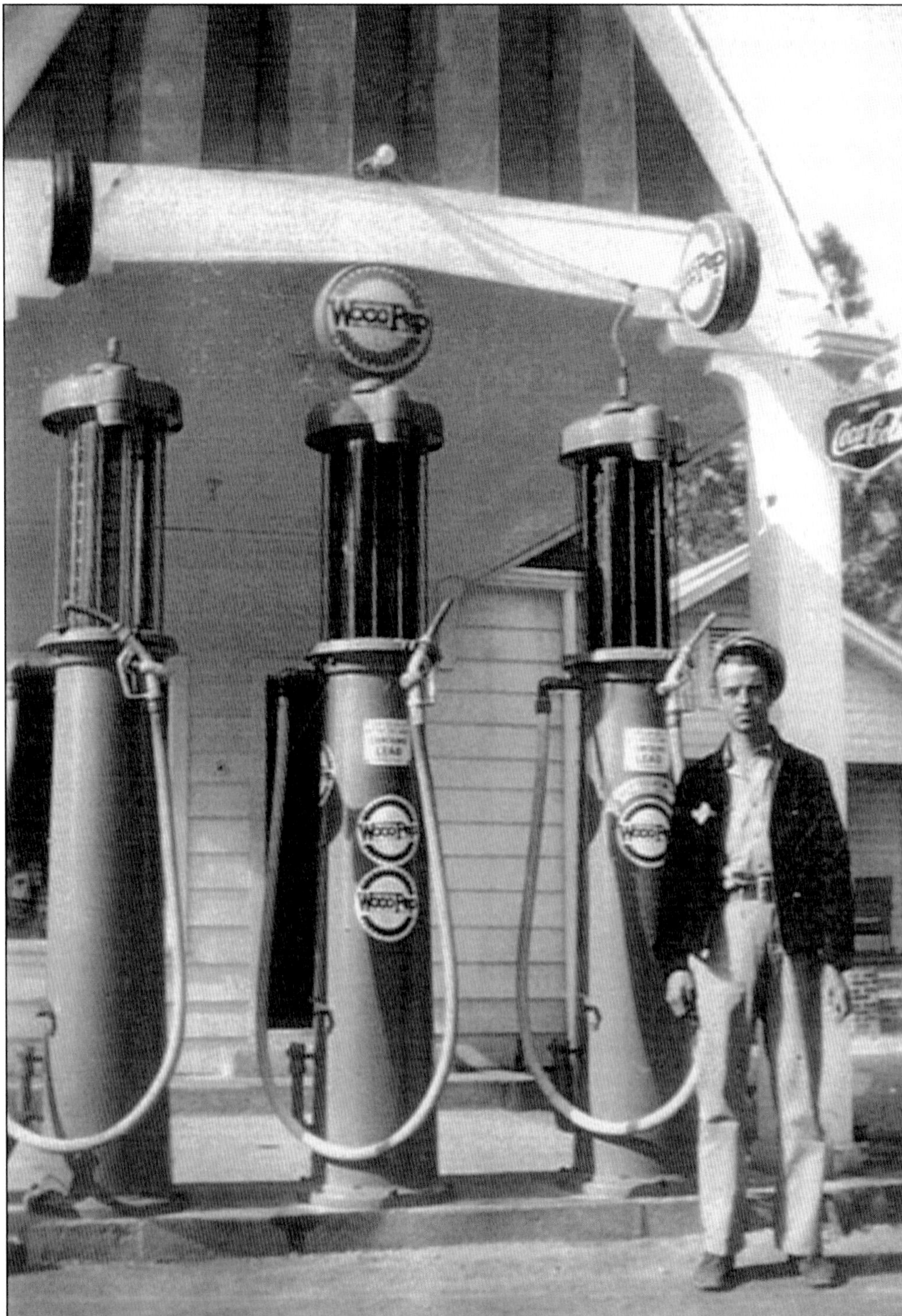

The Woco Pep station was located on Highway 31 in Pollard near the Little Escambia Creek bridges. Woco Pep was a fuel company out of Birmingham that was later sold to Pure Oil Company. (Courtesy of Jerry Fischer.)

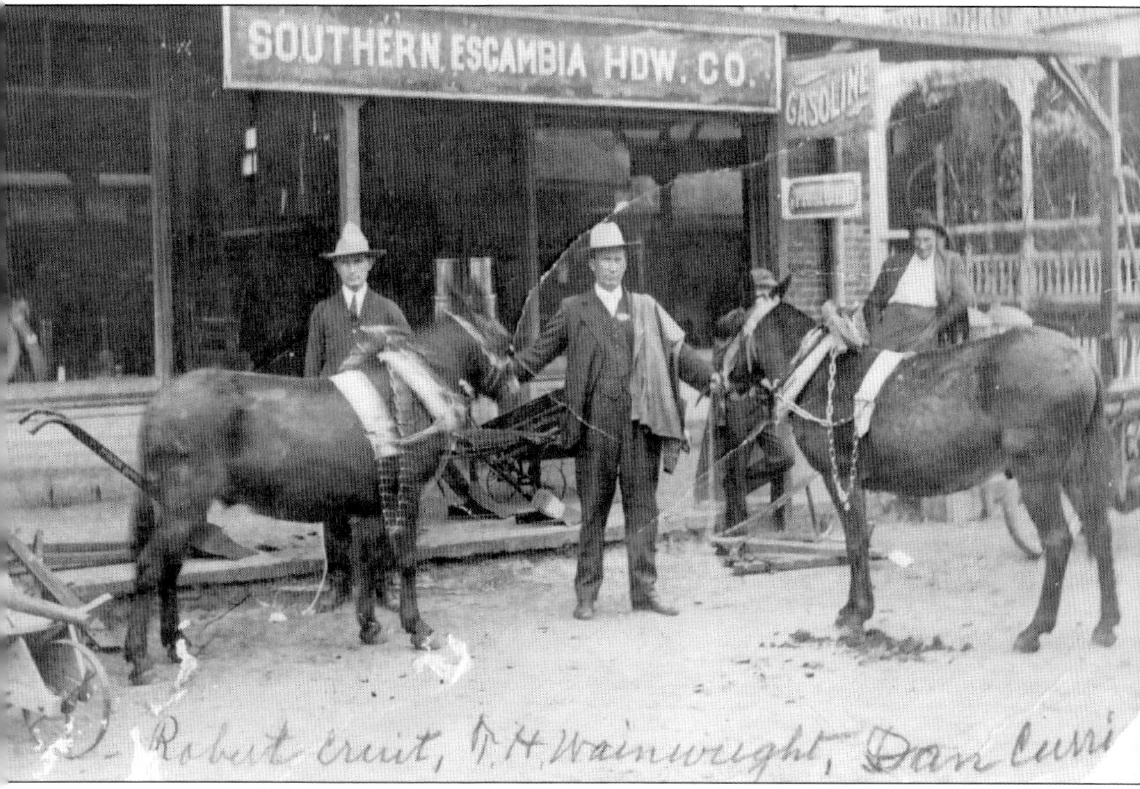

Robert Cruit (left), T.H. Wainwright (center), and Dan Currie stand outside of the Southern Escambia Hardware Company in Atmore, a business they opened in the early 1910s. Wainwright ran the hardware store, and the gas station was one of the first in the area. (Courtesy of Fritz Currie.)

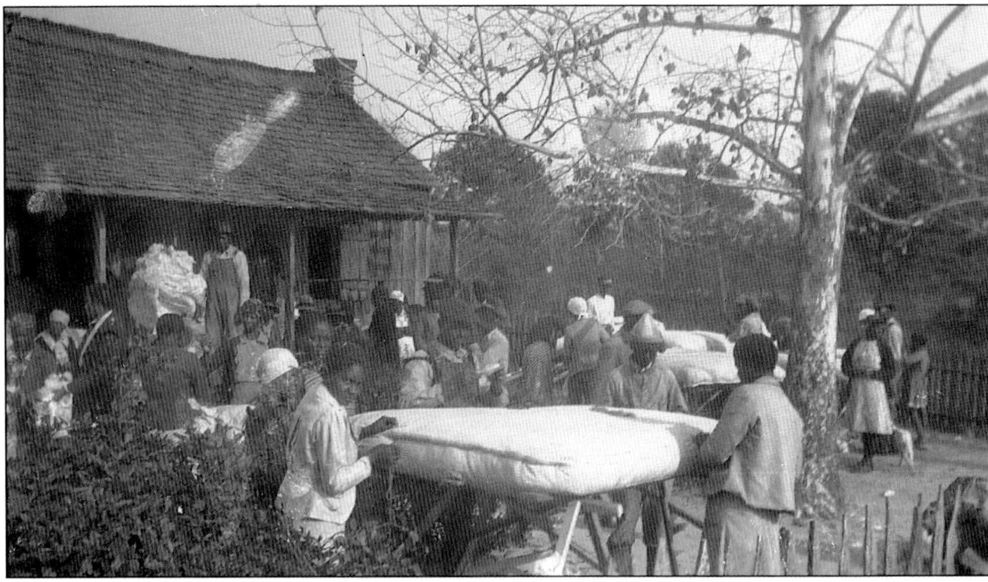

Farmhands are seen making mattresses using raw cotton that had been cleaned by hand. Escambia County had not been part of the plantation economy. In the early years, small farmers were responsible for the production of crops. After timber was cut, some lands were replanted, while other tracts were put to use growing crops, such as cotton, strawberry, and even tobacco for a short period. (Courtesy of D.W. McMillan Trust.)

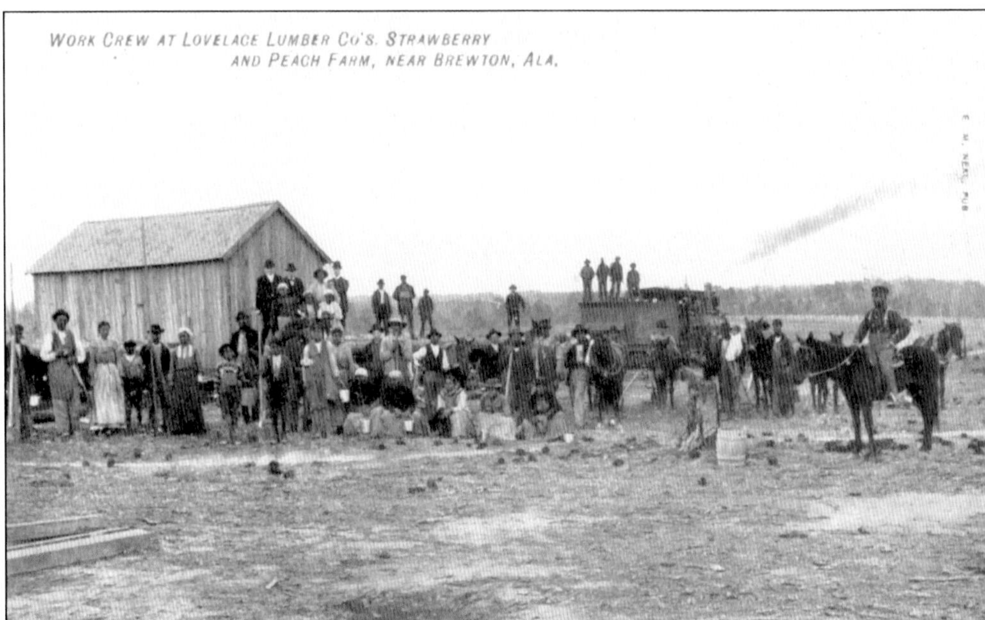

Workers at Lovelace Lumber Company's strawberry and peach farm near Brewton break for a photograph. The farm was originally planted on 200 acres of cutover pinelands before being expanded to an adjoining tract of 203 acres with plans of expanding to several thousand acres adjacent to the railroad and adding a canning factory. (Courtesy of D.W. McMillan Trust.)

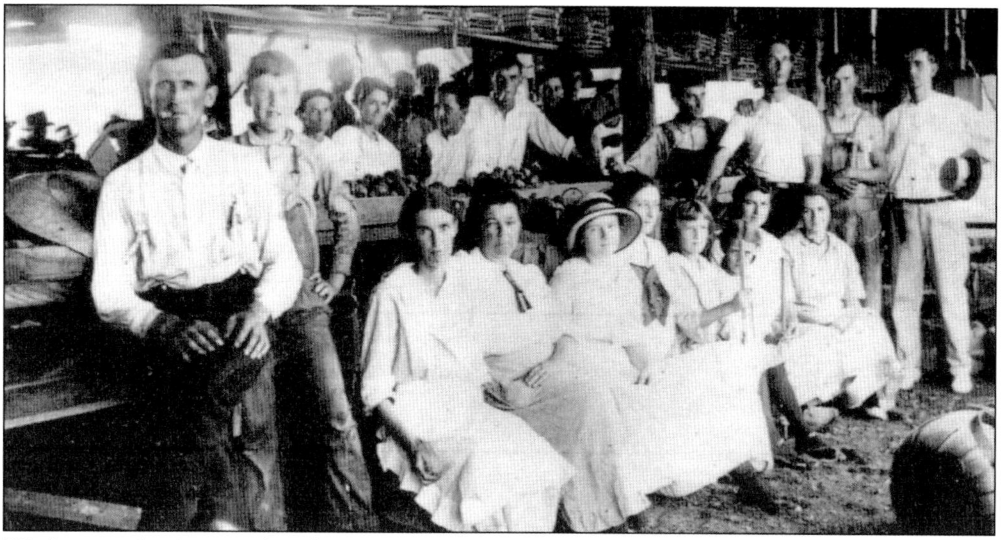

Workers at the fruit packinghouse on Dugout Lane in Canoe take a break for a photograph sometime in the 1920s. Here, strawberries, the major crop, were sorted, graded, and sent out of town by rail, along with Irish potatoes and sweet potatoes. Strawberry pickers earned 50¢ a day. (Courtesy of Carolyn Conn and Kevin McKinnley.)

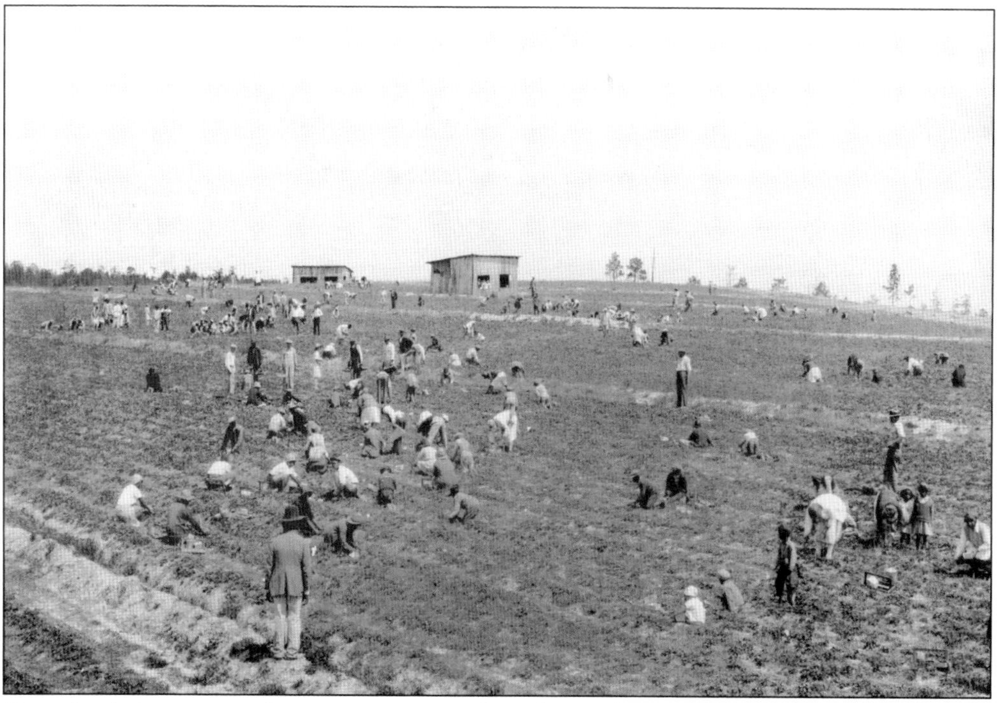

Once the timber was cut, this area was promoted as a fruit and vegetable growing area. Strawberries became a very big crop in the early 1900s. The area then became known for its blueberries after World War II. (Courtesy of D.W. McMillan Trust.)

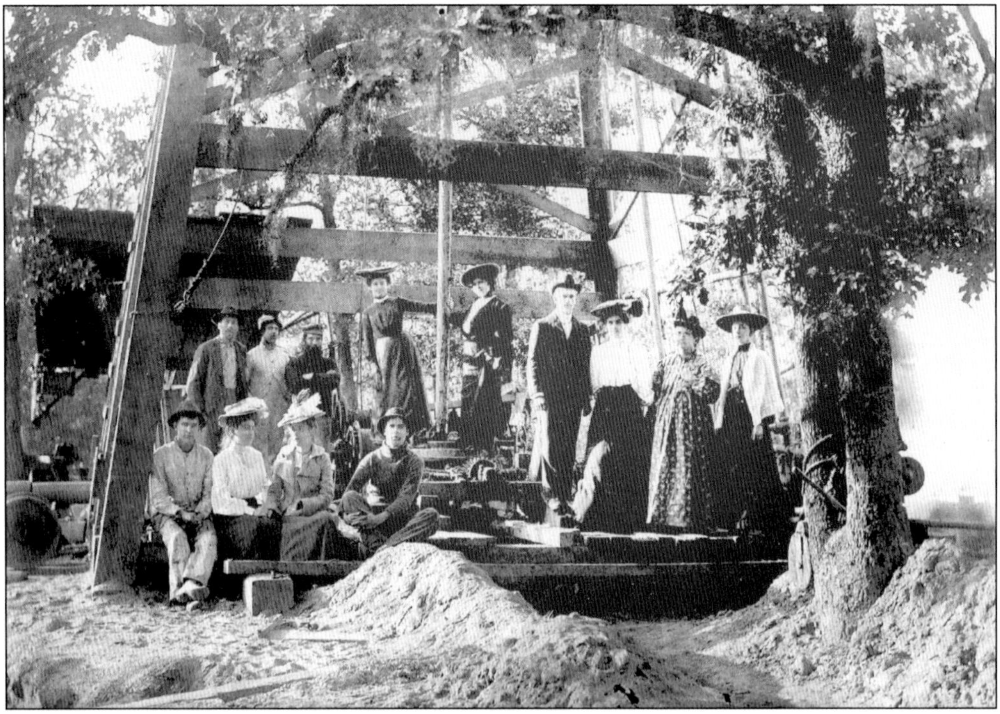

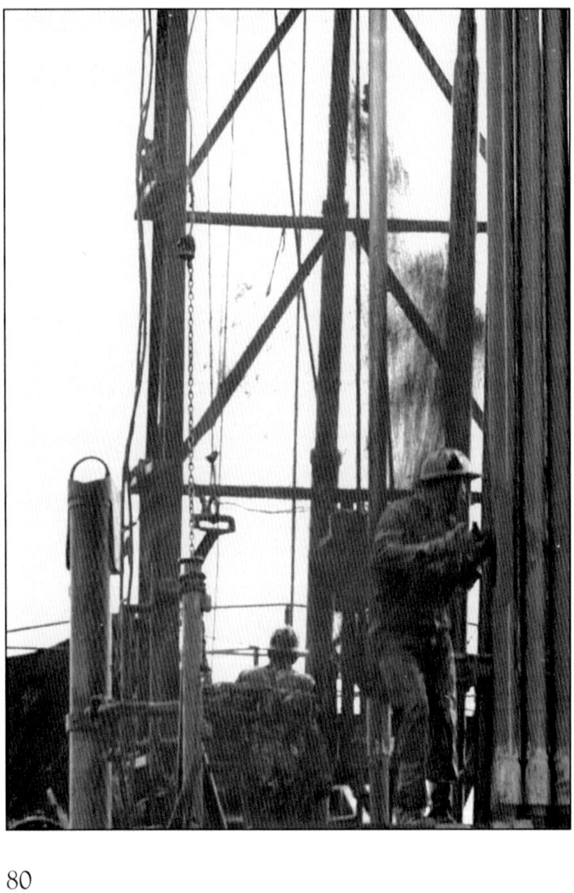

The first oil well in eastern Escambia County was drilled in 1902 on Conecuh River, six miles northeast of the small community of Roberts. Known as the Roberts well, it was drilled prior to the creation of the Alabama Oil and Gas Board. The well reached 1,640 feet into the earth, but no oil was found. Jack Canfield (second row, third from left) from Cleveland, Ohio, caused the well to be drilled. (Courtesy of Lydia Grimes.)

After many failed attempts, oil was finally discovered in Escambia County in 1952 by Humble Oil and Refining Company. At the Allen Moye well, oil, gas, and mud came rushing back to the surface—a real gusher. Today, this would not be allowed, as it is actually very unsafe and damaging to the environment. (Courtesy of D.W. McMillan Trust.)

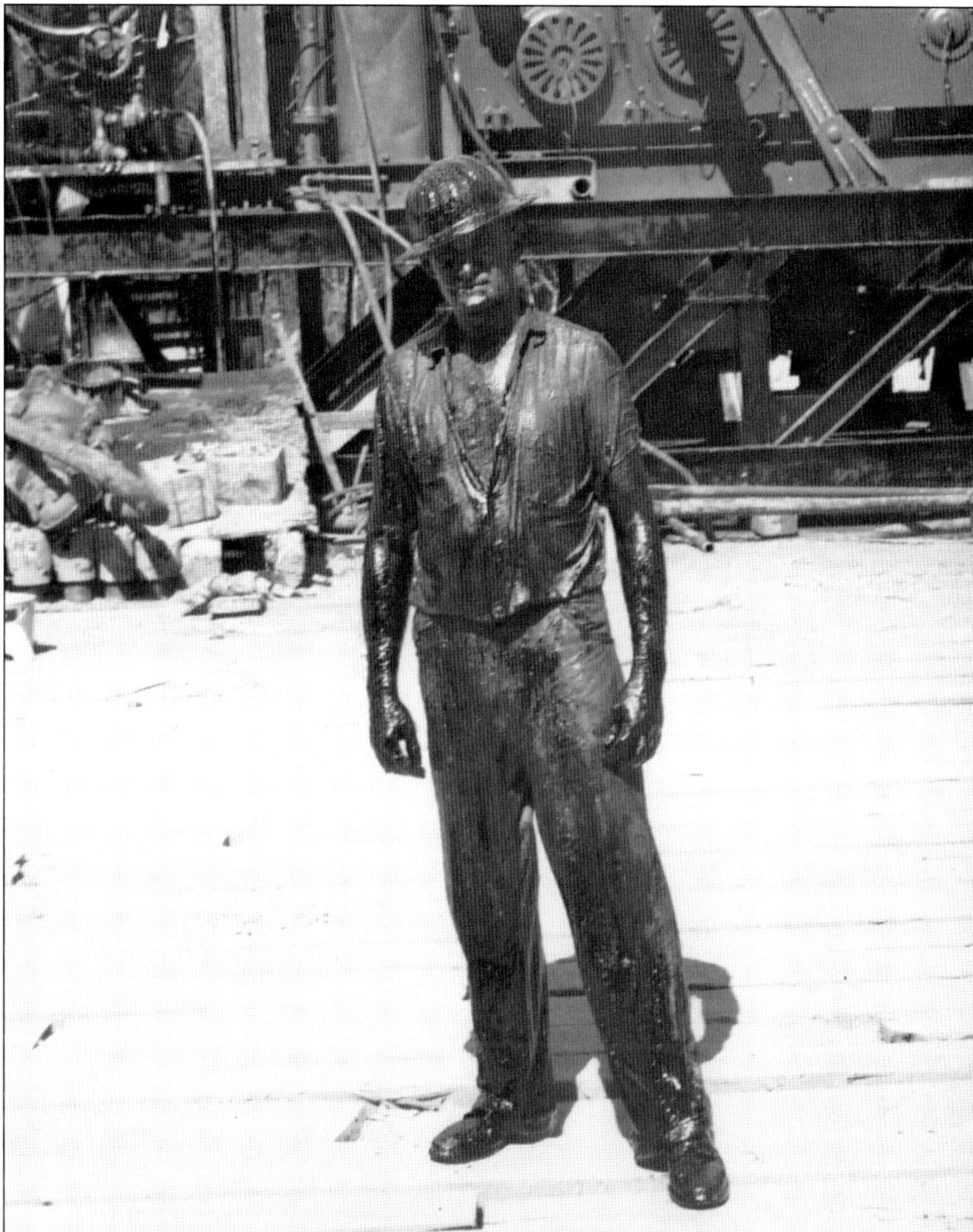

An oil worker stands covered with oil from head to toe at the Allen Moye well in 1952. Allen Moye was the discovery well at Pollard Oil Field in Escambia County. (Courtesy of D.W. McMillan Trust.)

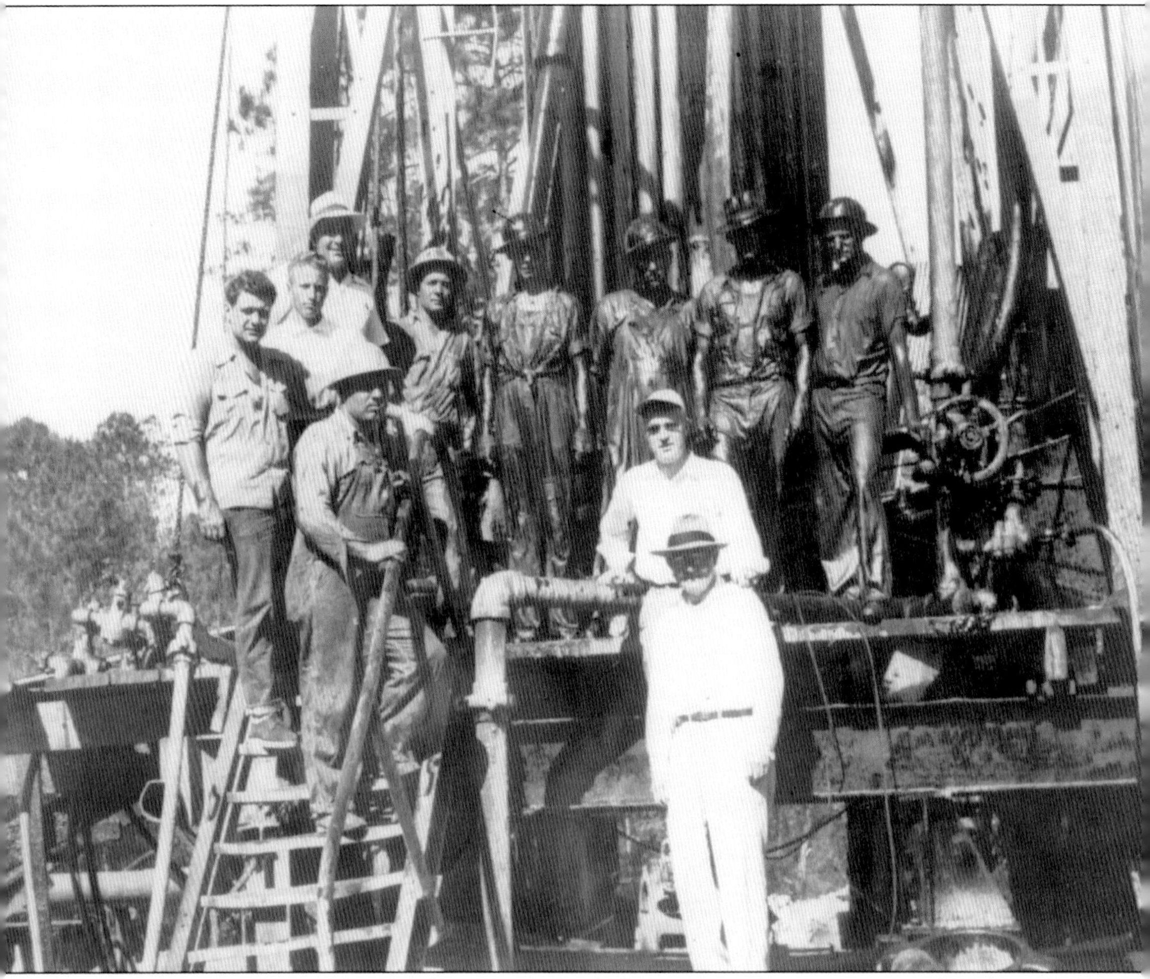

Ed Leigh McMillan (in front) is shown with the drilling crew at the A.W. Moye well of the Pollard Oil Field in 1952. The oil field was the first discovery in Escambia after many attempts. This gusher created a flurry of excitement that led to many more discoveries in the area. (Courtesy of D.W. McMillan Trust.)

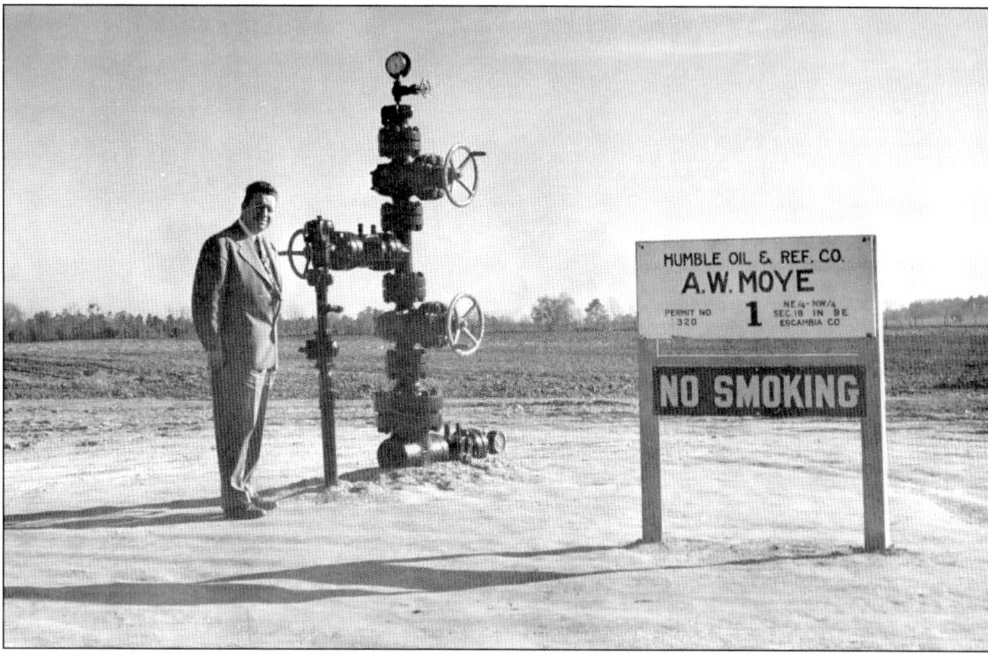

After the discovery well at Pollard had been connected to the gathering system, state geologist Dr. Walter B. Jones made a visit to Escambia County. Dr. Jones became well known in the area and made many lifelong friends from his frequent visits. (Courtesy of D.W. McMillan Trust.)

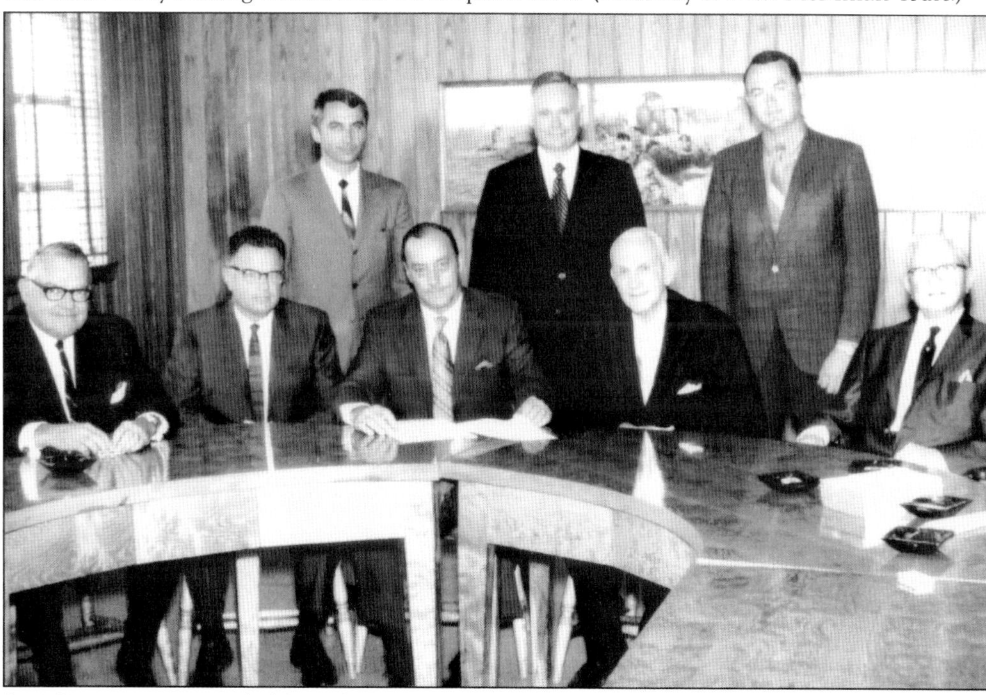
In 1970, a meeting was held on the occasion of an oil and gas lease signing between T.R. Miller Mill Company (lessor) and Louisiana Land and Exploration Company of New Orleans (lessee). Escambia County has been and continues to be a place of interesting geology and lucrative oil and gas deposits. (Courtesy of D.W. McMillan Trust.)

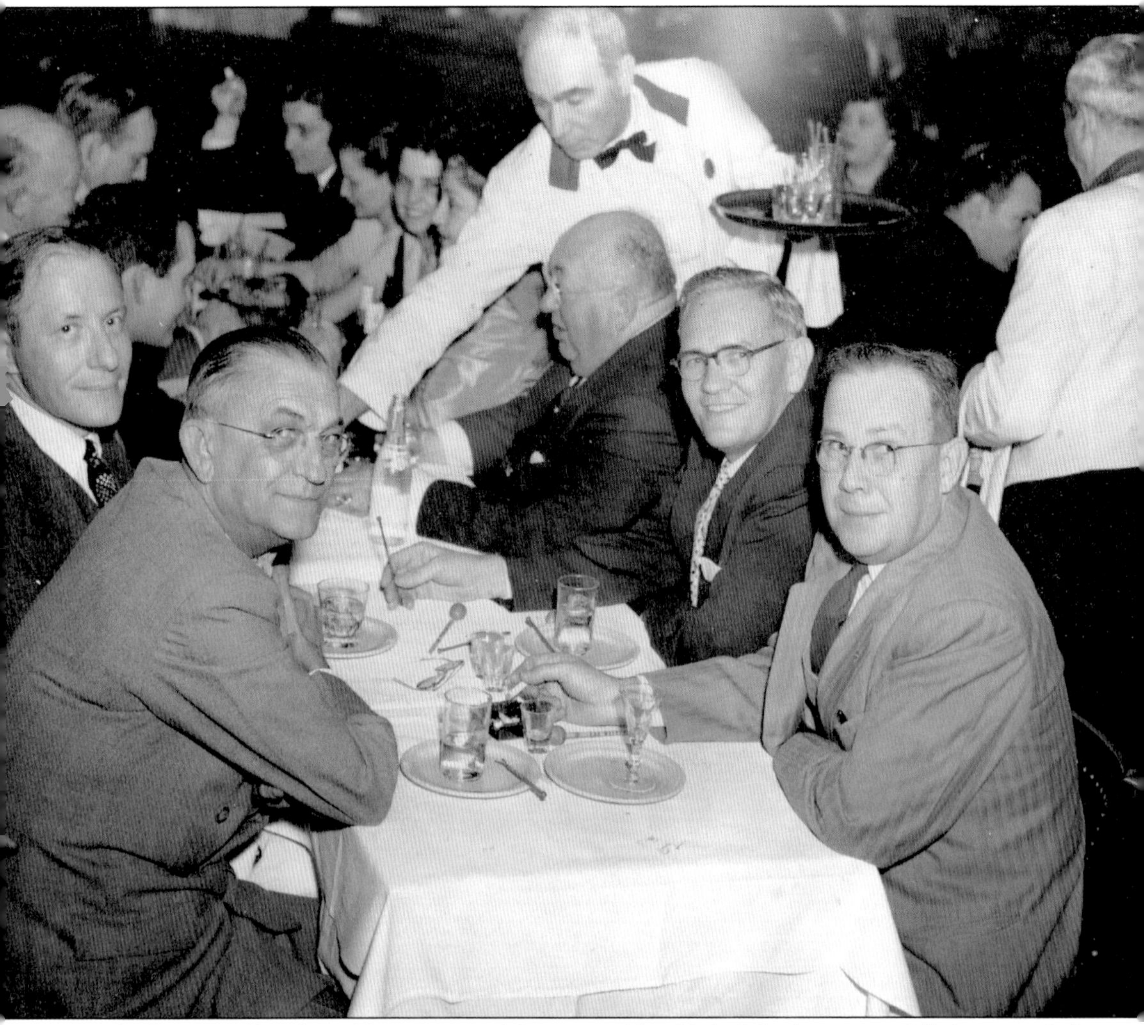

The Brewton Chamber of Commerce celebrates in a New York club after successfully bringing Judy Bond, a ladies' garment maker, to town in 1961. Milliard Rothenberg, back left, was the co-owner of Judy Bond. Sitting next to him, William E. Brooks was the owner of local radio station WEBJ and a member of the chamber. Jack Rothenberg, back right, was a co-owner of Judy Bond, and next to him is Ed Mac Lovelace, the president of the chamber. Jim Smith, front right, was the president of the Bank of Brewton and a chamber member. (Courtesy of Yank Lovelace.)

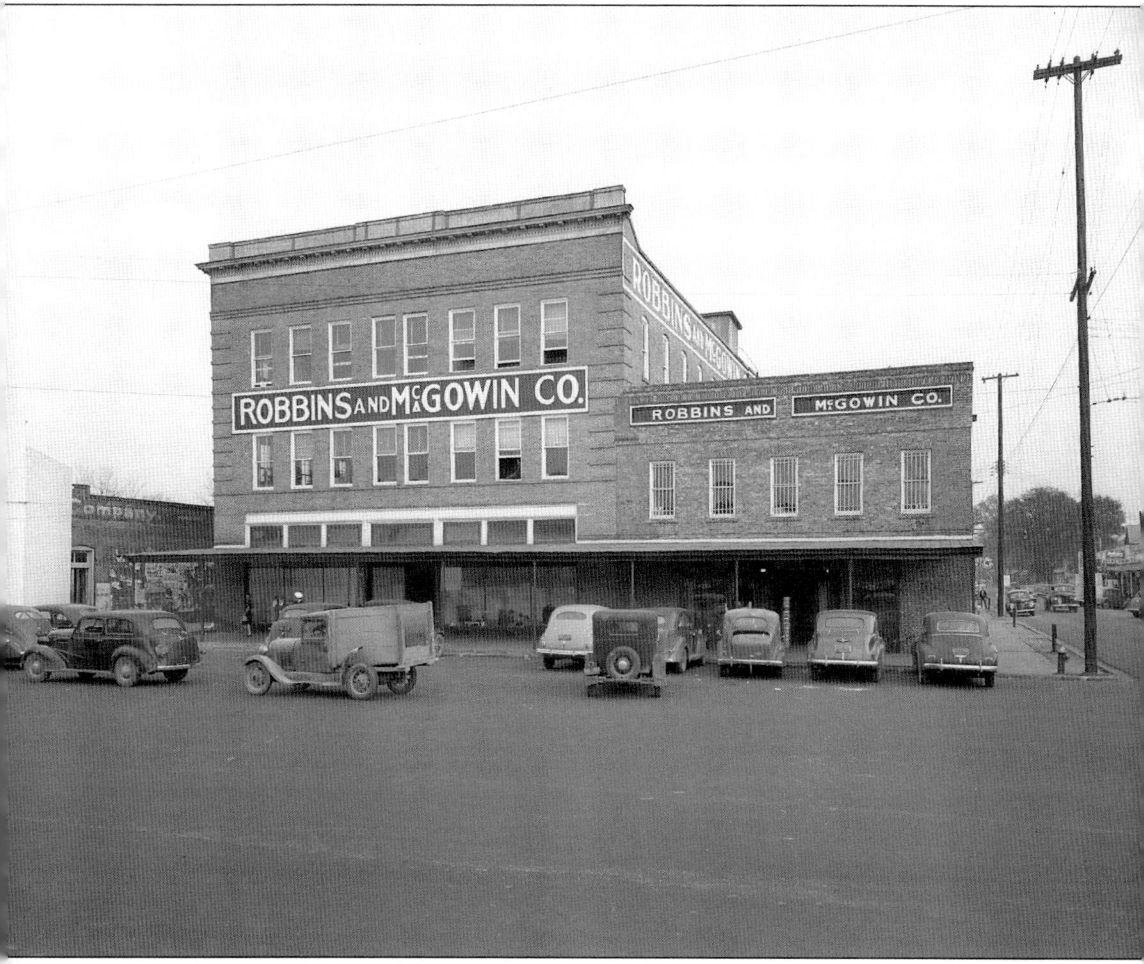

Cars are parked outside the Robbins and McGowin Company on Highway 31 in Brewton. In 1911, the Robbins and McGowin Company was advertised as the largest general store in Alabama in the *Brewton Standard* newspaper. The store had a mail-order catalog. (Courtesy of D.W. McMillan Trust.)

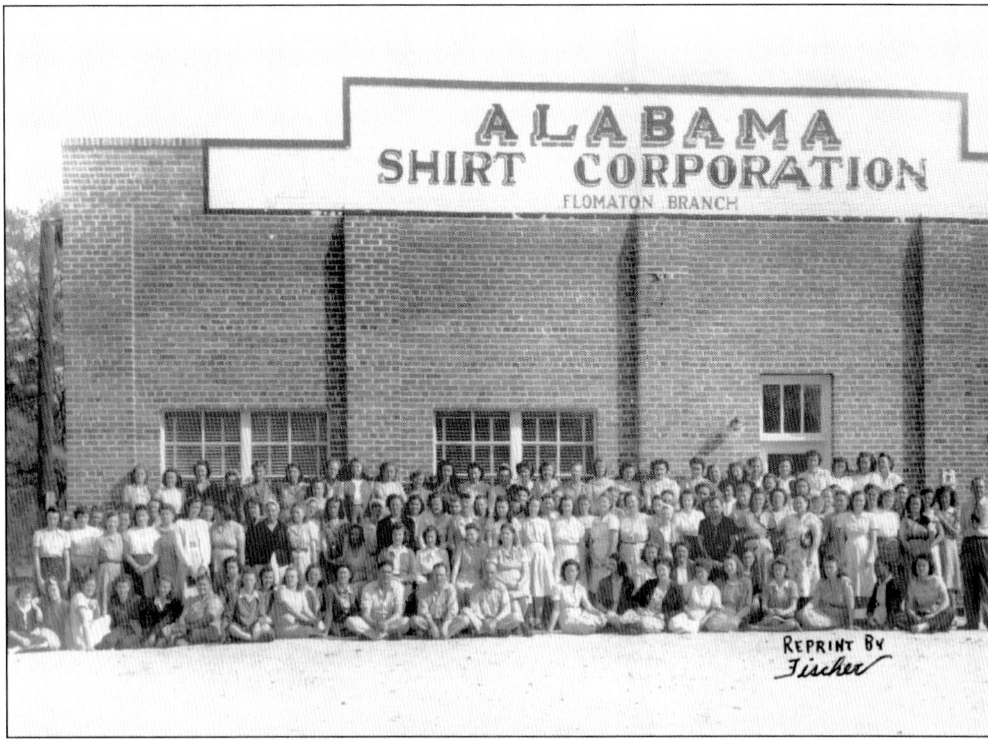

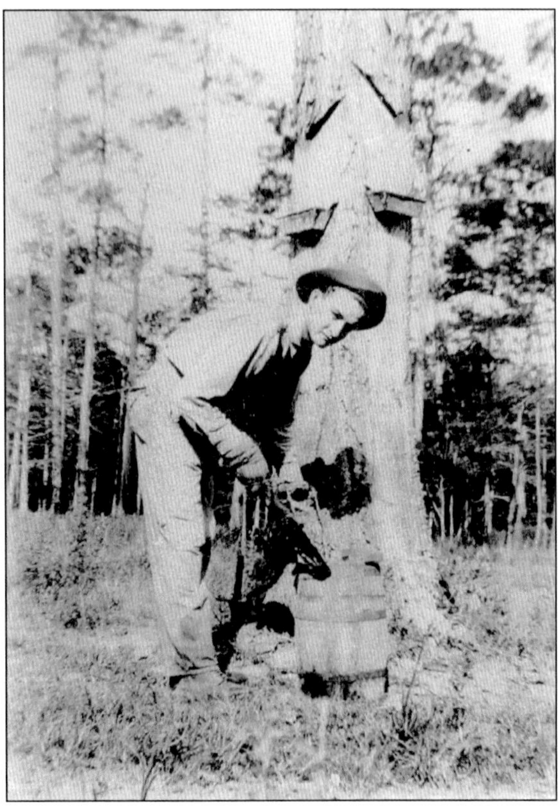

Workers at the Alabama Shirt Corporation in Flomaton break for a photograph. After World War II, the textile industry in Alabama became a large employer across the state. Later, cut-and-sew operations became very important to the economy of many small towns. The Alabama Shirt Corporation employed many female workers. (Courtesy of Jerry Fischer.)

A worker collects pine rosin into buckets for the production of naval stores. The pattern of angled cuts into the tree that let the sap drip into cups was known as "cat face." (Courtesy of Jerry Fischer.)

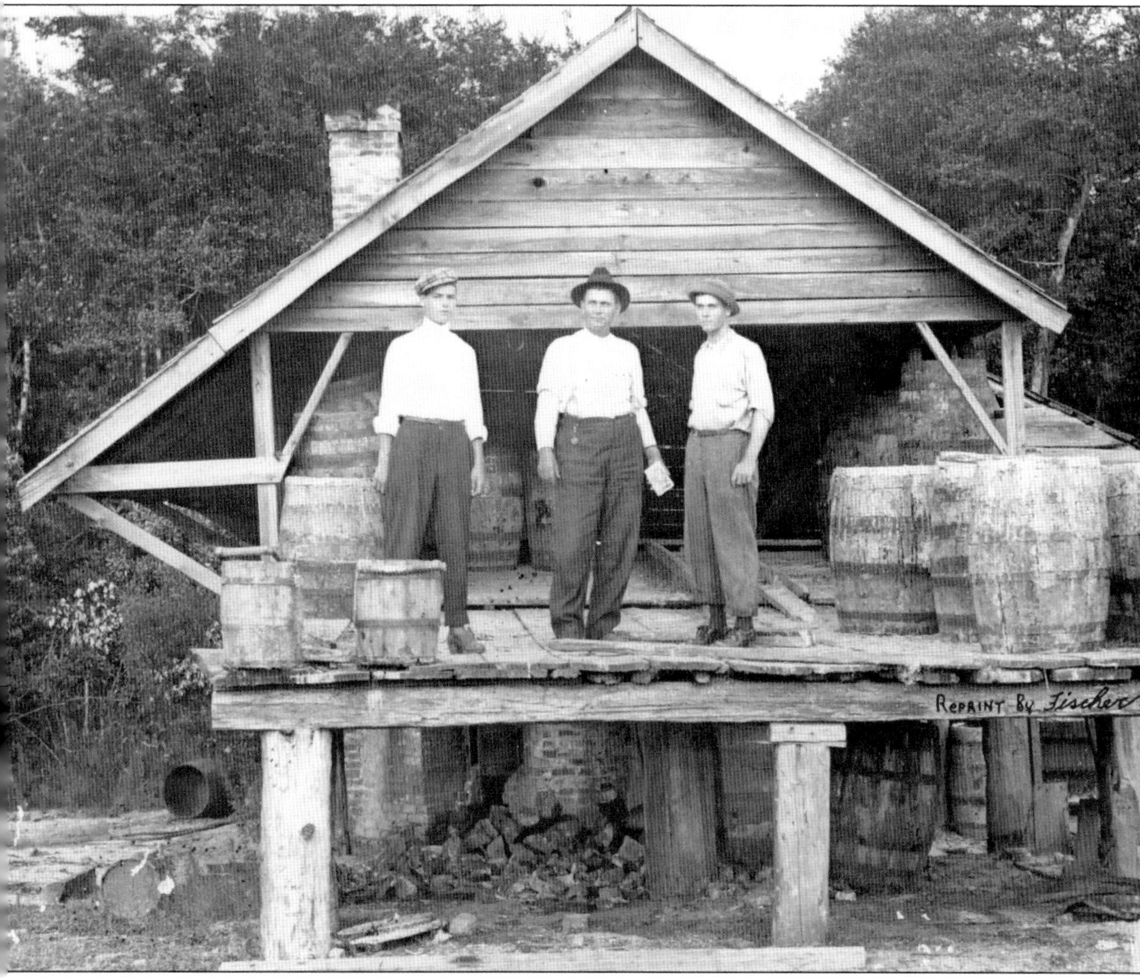

The Mize family stands with their small turpentine operation at Pineview. Turpentine was processed from small buckets into large barrels before being shipped out to make naval stores, such as paints and varnishes. The platform is raised to allow transfer from wagons. (Courtesy of Jerry Fischer.)

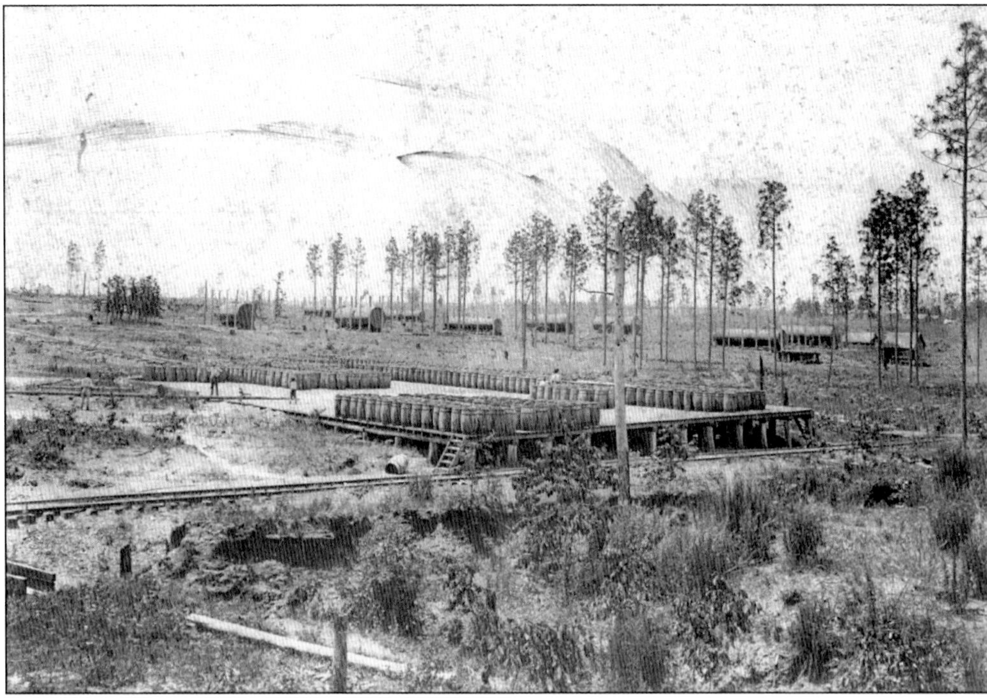

Numerous barrels of naval stores are arranged on a platform, known as a rail siding, to be exported. The term "naval stores" originally referred to any resin-based products, such as turpentine, rosin, resin, and tar, used in the making of wooden ships. The term eventually became applied to any product containing pinesap. (Courtesy of Jerry Fischer.)

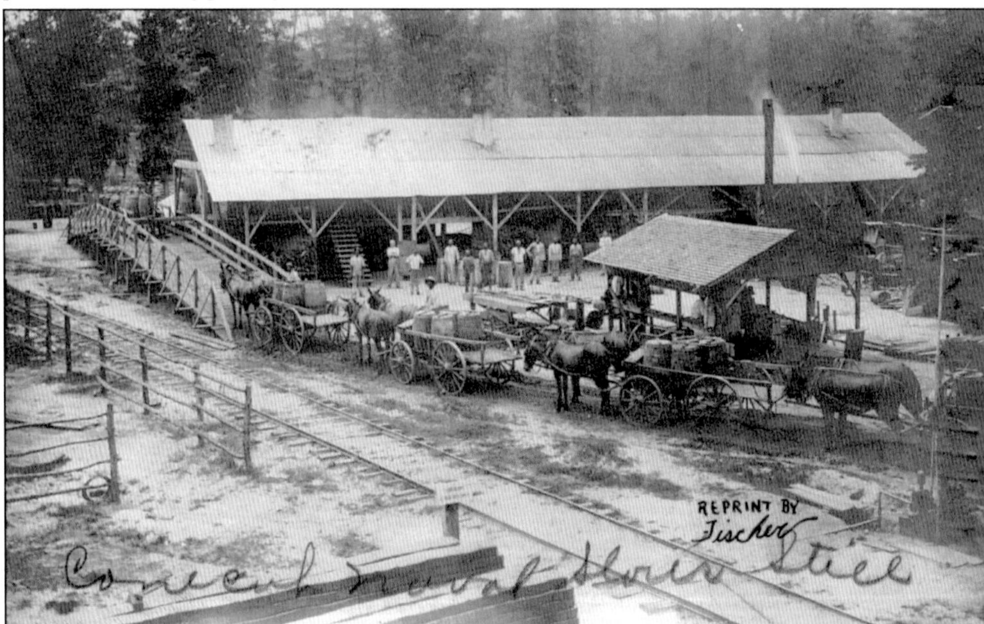

Wagons carry barrels of turpentine waiting to be unloaded at Hammac, a small community north of Flomaton. As the timber market matured, many landowners also realized the added benefit of extracting turpentine from the tree. (Courtesy of Jerry Fischer.)

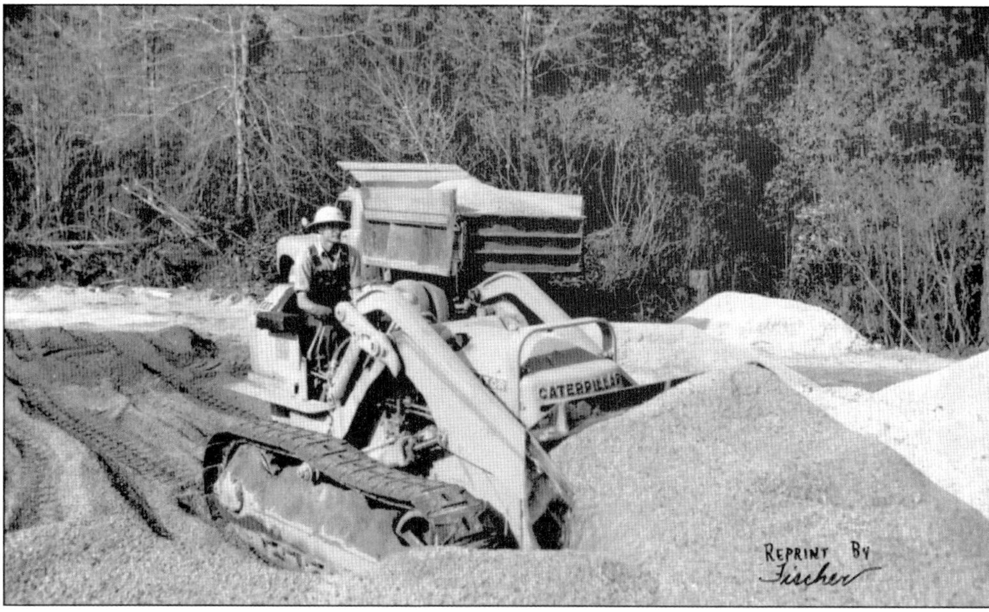

Jeff Emmons loads gravel in 1953 in the early days of Fore Sand and Gravel, an operation owned by Dick Fore in Flomaton. Escambia County is blessed with many natural resources, including sand, gravel, and red clay. (Courtesy of Jerry Fischer.)

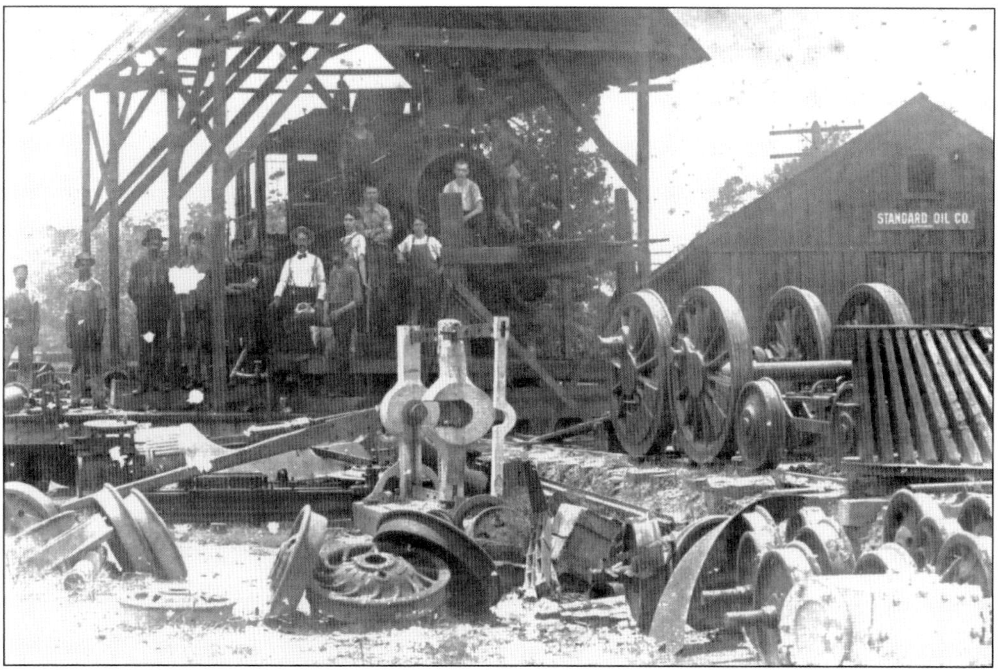

In 1910, workers overhaul a steam locomotive at Brewton Iron Works, a wholly owned subsidiary of Lovelace Lumber Company. W.F. Wilson became 25 percent owner in 1910 and later the sole owner in 1922. Wilson is pictured in the middle in the white shirt and bow tie. Sawmills had to provide many services for themselves and for their workers. Most mills had general stores, some had ice plants, and some had machine shops to fix and repair the heavy machines and locomotives. (Courtesy of D.W. McMillan Trust.)

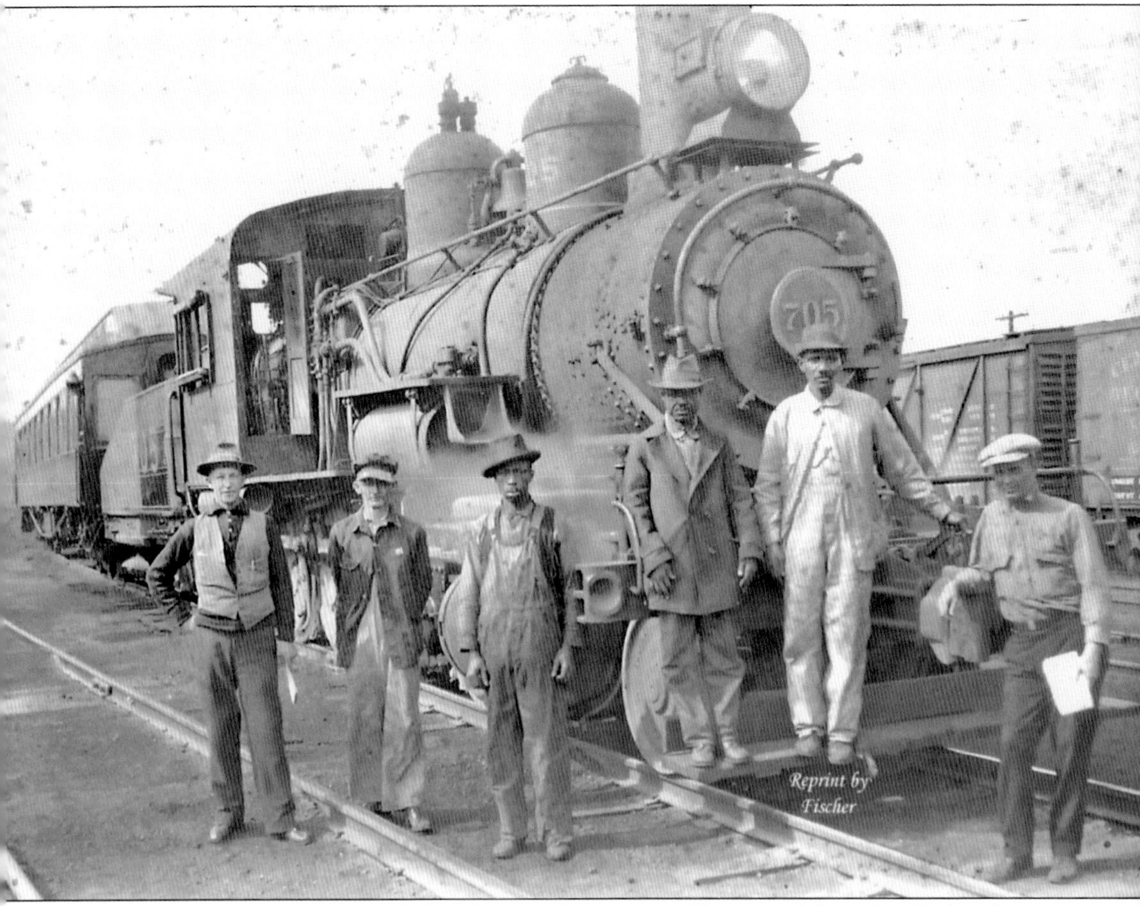

An L&N Railroad crew stands beside engine No. 705 at Flomaton. The L&N Railroad was one of the larger employers in town and provided some of the better jobs in the area. (Courtesy of Jerry Fischer.)

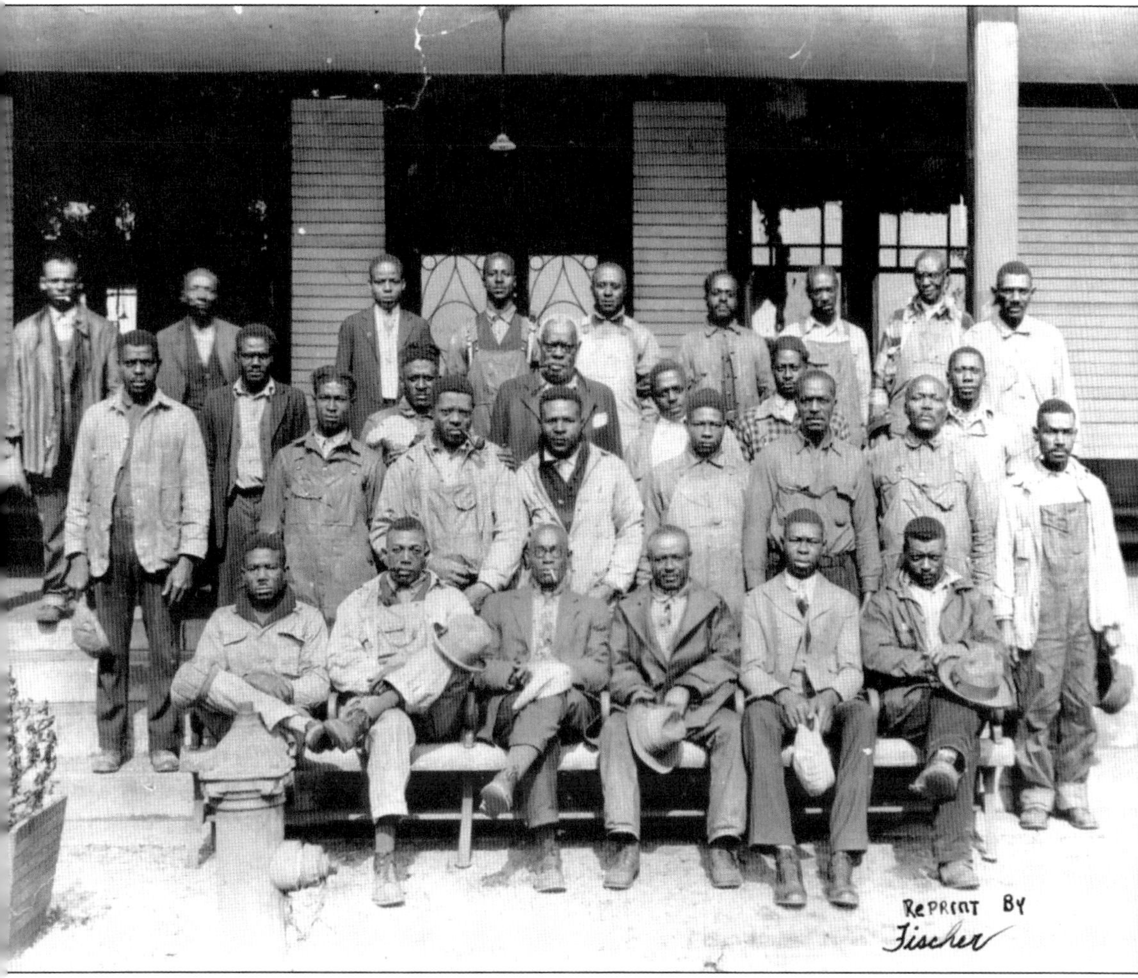

L&N workers gather for a photograph at the Flomaton depot around 1930. (Courtesy of Jerry Fischer.)

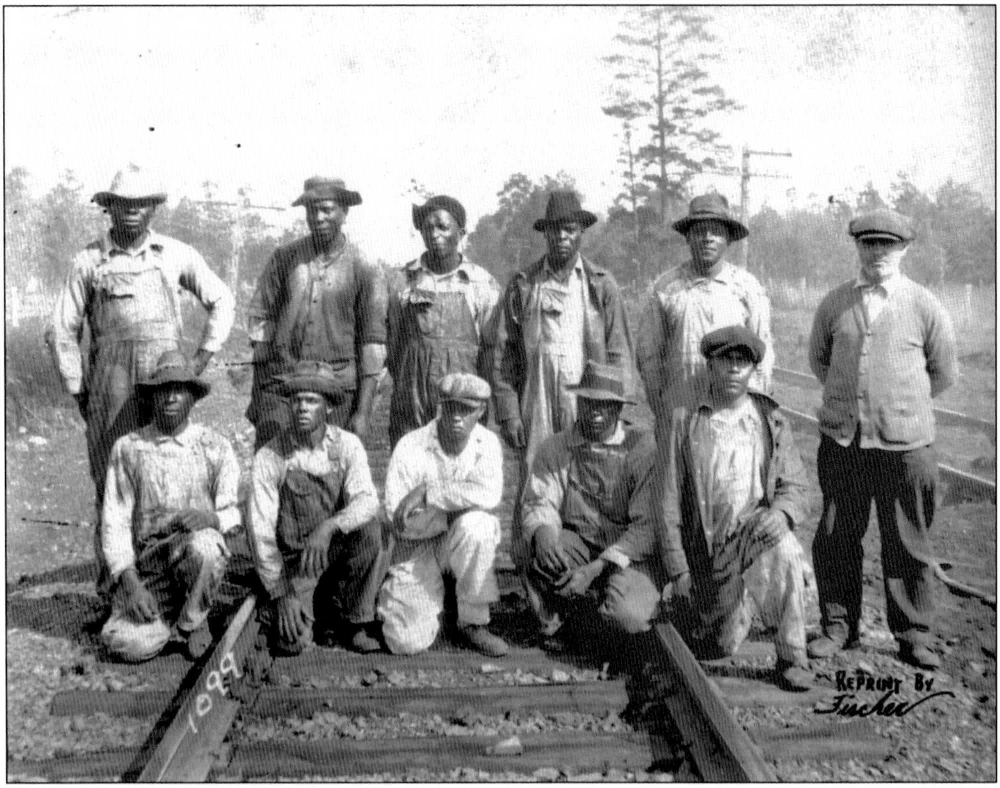

This L&N Railroad section crew's job was to maintain the tracks. The crew boss (far right) was named Driscoll. A big day for the section crew was the when the pay train (some called it the sugar train) arrived. The crew was paid in silver coins, and in a short time everyone in town would be jingling new silver dollars. (Courtesy of Jerry Fischer.)

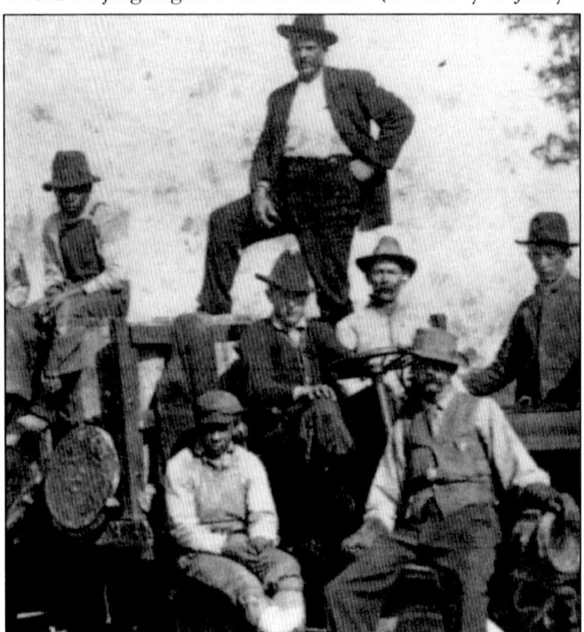

A group of men hauls turpentine in Canoe in 1914. Turpentine was gathered using the same process used in collecting maple sap with tin turpentine "cups" attached to the bottom of pine trees. Old tins were not discarded but were pounded flat and used as roofing shingles. (Courtesy of Carolyn Conn and Kevin McKinley.)

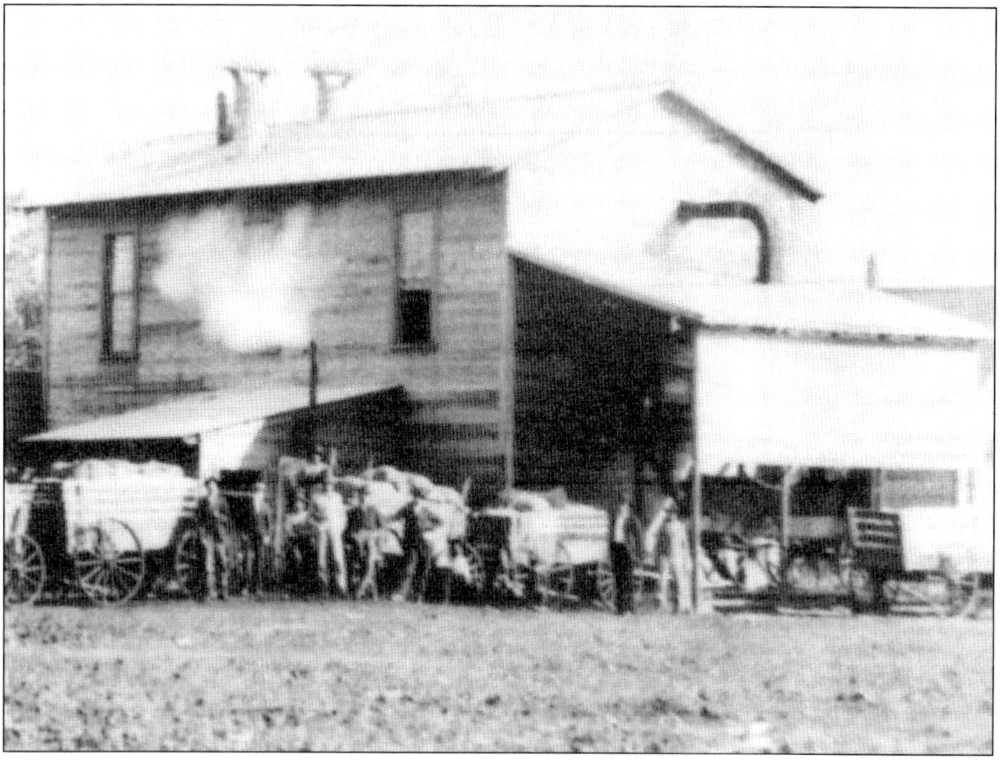

Workers stand outside of a 1909 company that was a cotton gin, turpentine operation, and grits mill all in one in Canoe. The company was located on Baker Street just south of Highway 31 and shipped its goods out by rail. (Courtesy of Carolyn Conn and Kevin McKinnley.)

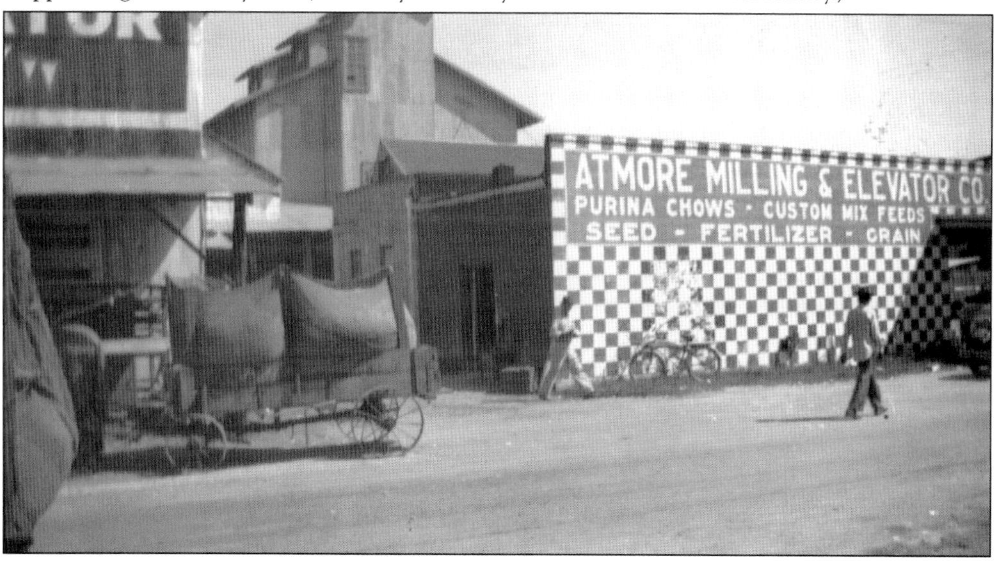

The Atmore Milling and Elevator Company was the center of the Atmore Agriculture District. Its three gins handled the cotton, while a gristmill turned out grits. Large cucumber, potato, and corn sheds were nearby, which attracted migrant workers looking for seasonal work. Women sorted the cucumbers, which were sent for pickling. A potato-chip company executive once told workers that the best potato chips were made from Atmore's potatoes. (Courtesy of Fritz Currie.)

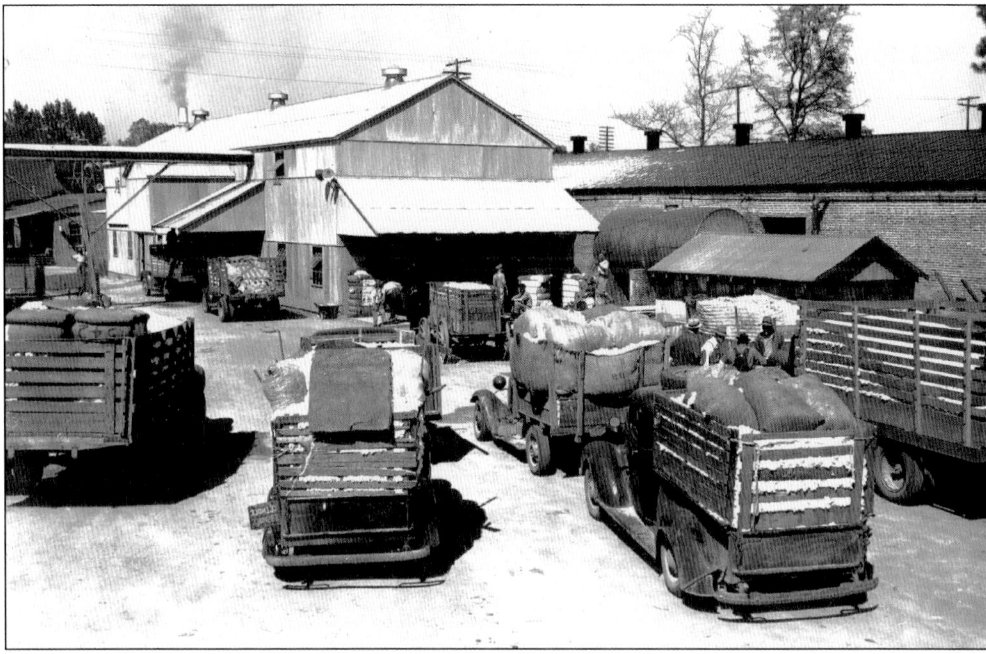

Cotton was king in the 1950s when it arrived on a Saturday at Currie's Gin. Currie's was one of three local gins, along with Atmore Truckers and Frank Currie's. The ginning companies held a friendly contest to gin the first bale of cotton. The winner received a cash prize and was honored in a special ceremony. (Courtesy of Fritz Currie.)

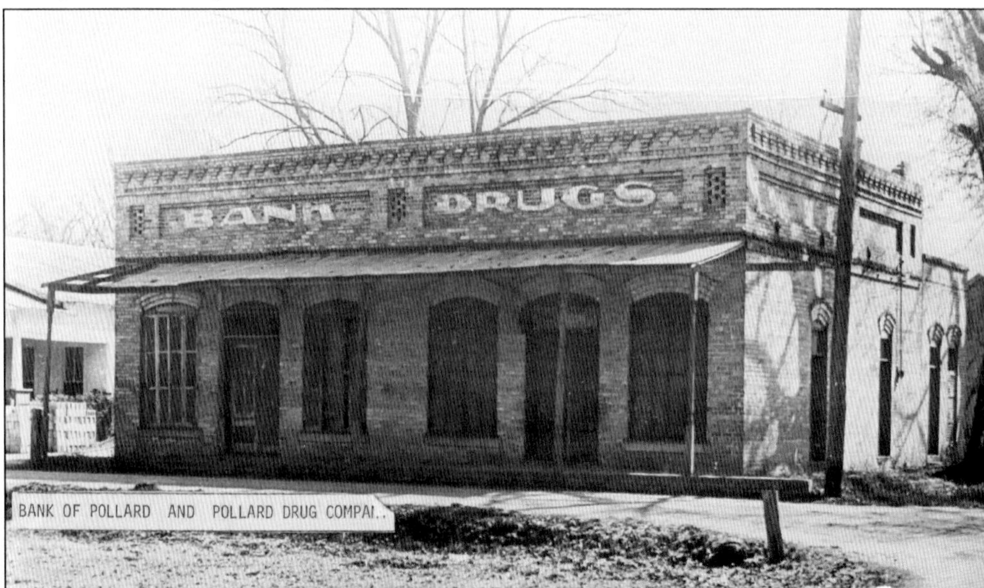

Pollard was the junction of three railroads: the Alabama & Florida from Montgomery, the Florida & Alabama from Pensacola, and the Great Northern from Mobile. The Bank of Pollard was organized in 1907 with Martin Lindsey as president, Sam Pouncey as cashier, and H.D. Finlay and Sam Pouncey as directors. The Bank of Pollard and Pollard Drugstore were victims of the flood in 1929 and the Depression that followed. (Courtesy of Lydia Grimes.)

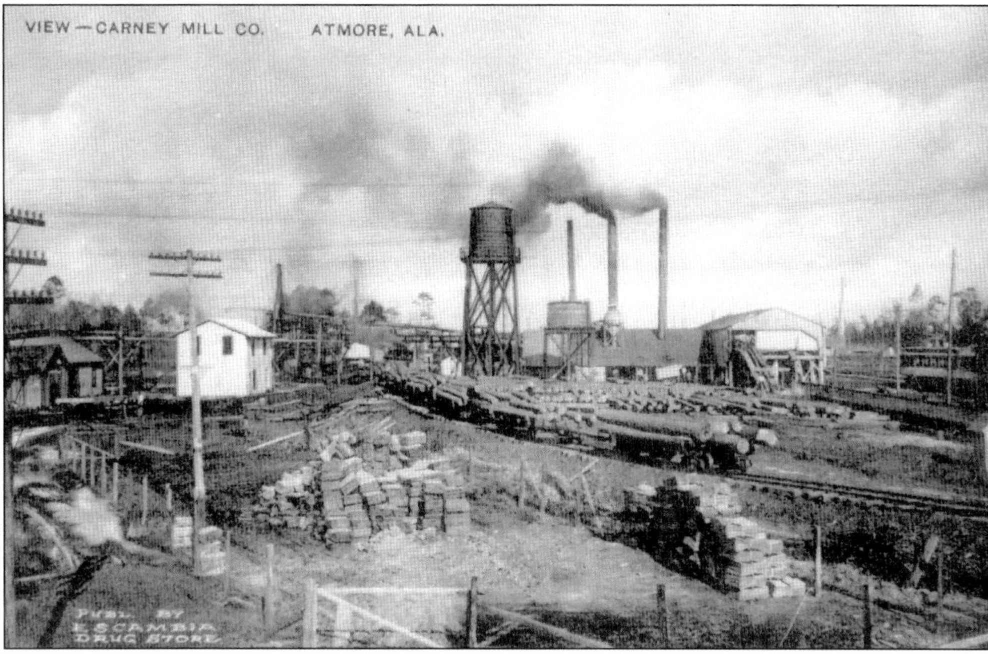

Carney Mill Company was established at William Station in Atmore in 1881 by William Carney. He erected the first mill on Indian Pond just east of town. It was later moved closer to town and grew to be one of the largest mills in Escambia County. It closed in the autumn of 1930 because of the Depression. (Courtesy of Yank Lovelace.)

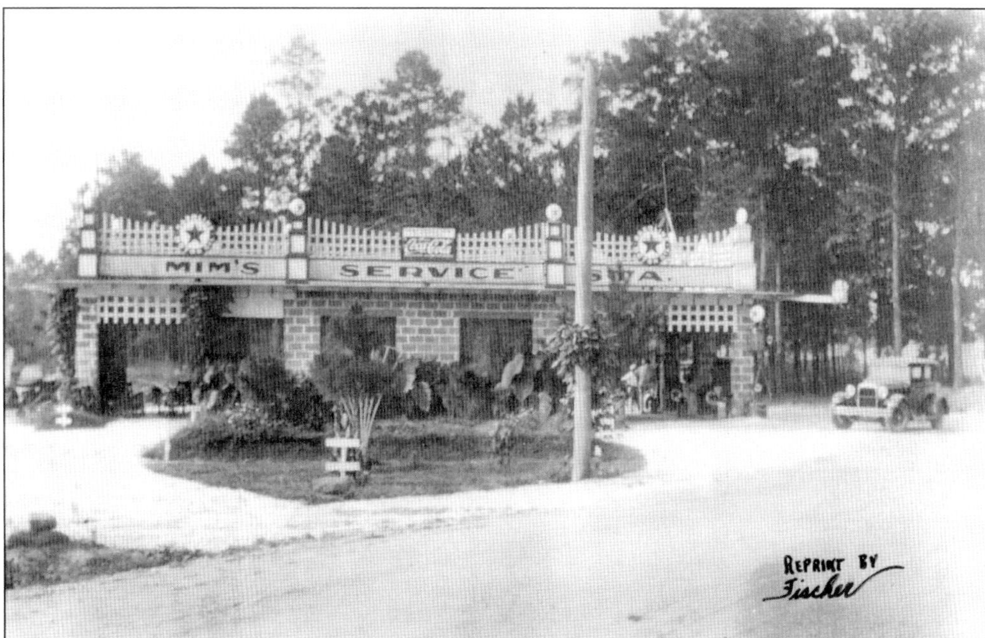

Mim's Service Station was located in Flomaton at the intersection of US Highways 31 and 29 to Pensacola, and its landscaping offered a real personal touch. Most of the early service stations in the area were locally owned, and they provided a lot of personal service to motorists. (Courtesy of Jerry Fischer.)

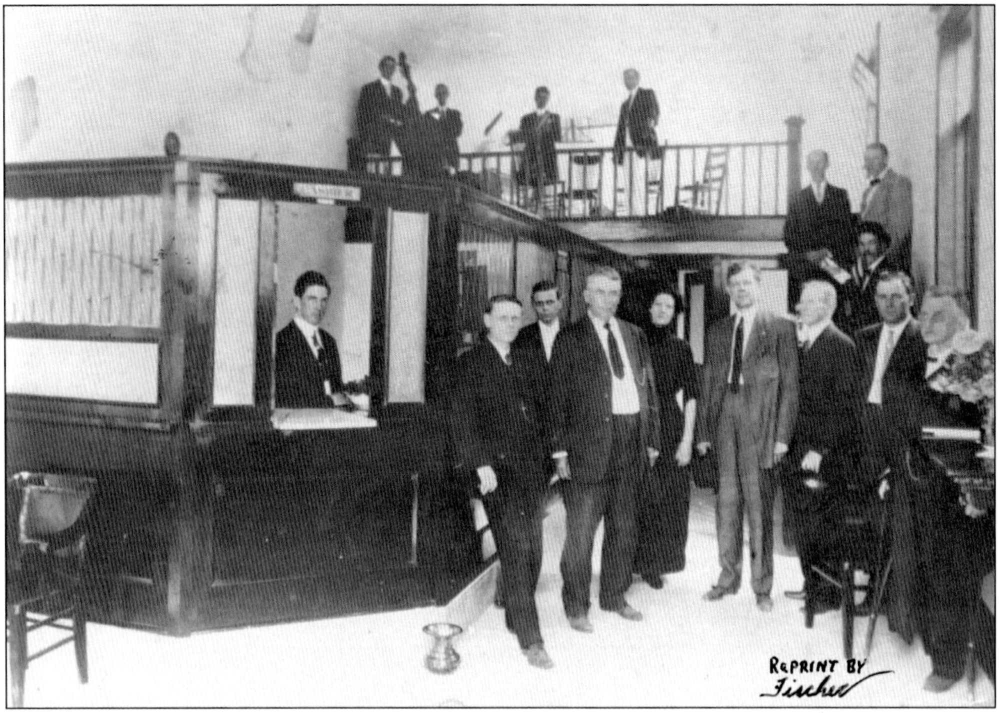

Employees and stockholders celebrate the grand opening of the Escambia County Bank in 1924 in downtown Flomaton as a string band plays on the balcony. (Courtesy of Jerry Fischer.)

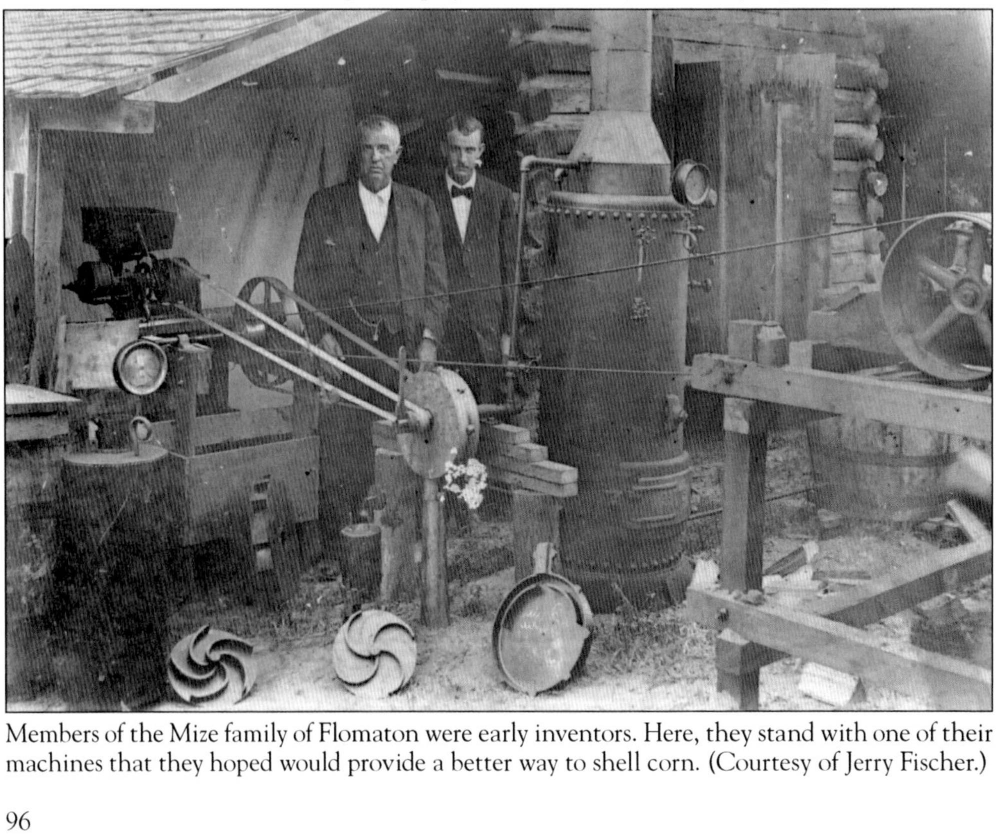

Members of the Mize family of Flomaton were early inventors. Here, they stand with one of their machines that they hoped would provide a better way to shell corn. (Courtesy of Jerry Fischer.)

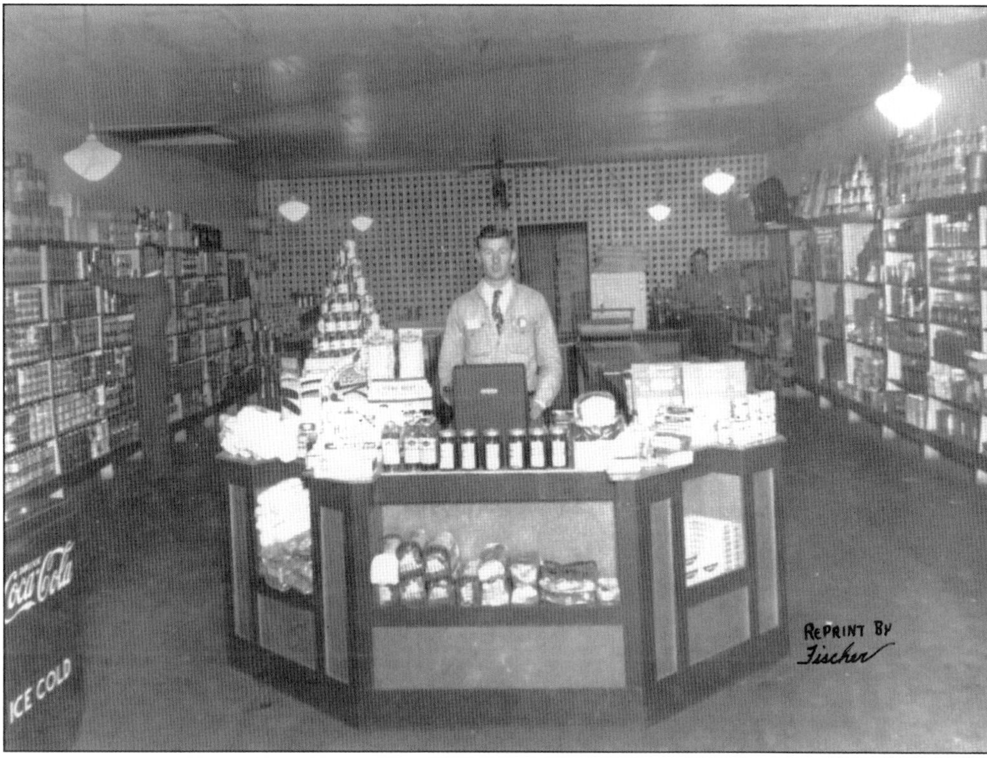

Lloyd Harper stands at the cash register of his Flomaton grocery store. He spent most of his life, about 50 years, working in the store. At one time, there were as many as five grocery stores in town. (Courtesy of Jerry Fischer.)

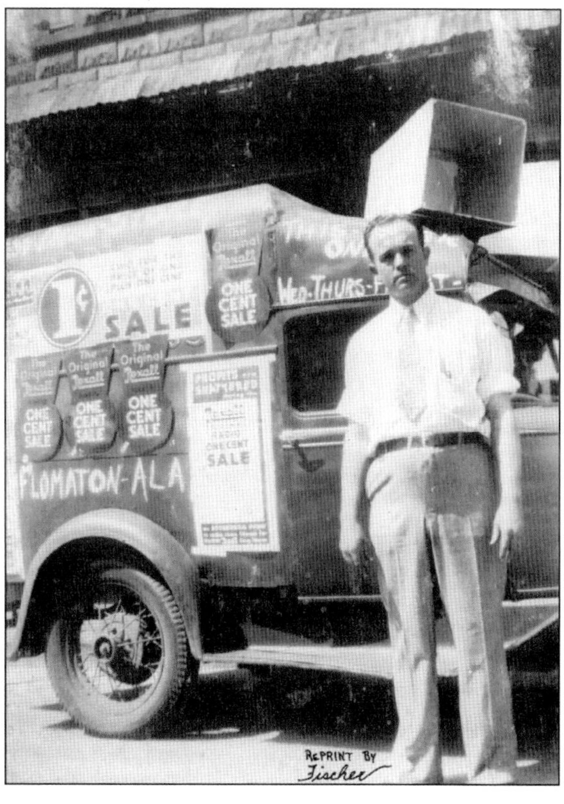

Lloyd Harper stands with his so-called "rolling store," which was one of the first to travel the county roads. (Courtesy of Jerry Fischer.)

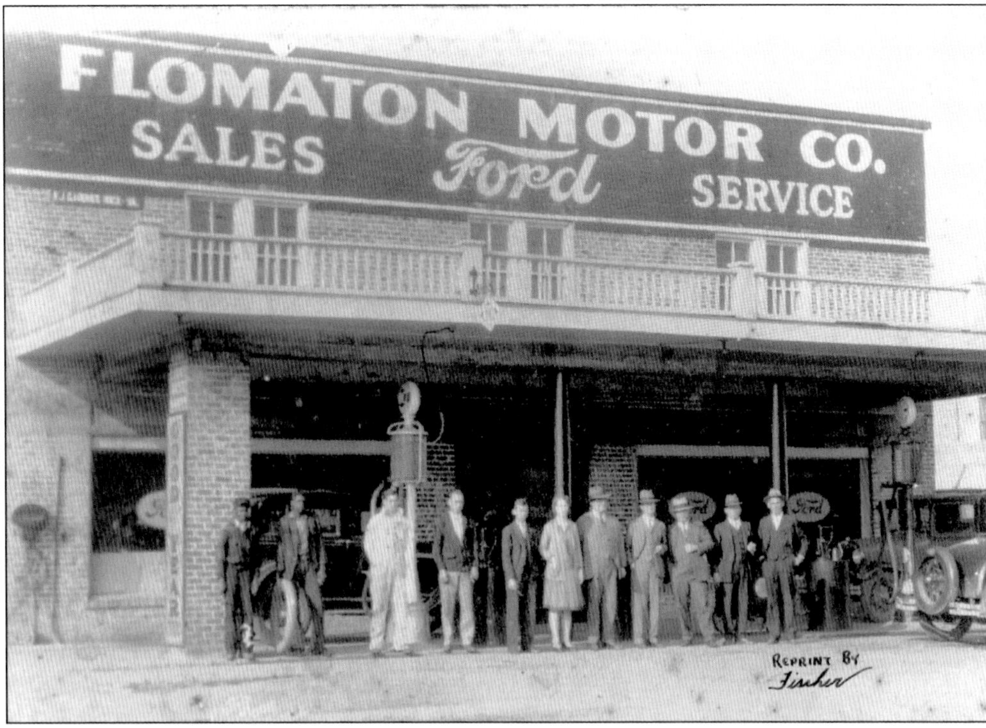

The Flomaton Motor Company was owned and operated by Alton Ray Beatha and sold gasoline. Previously, the building had been a bottling plant; it still stands to this day. (Courtesy of Jerry Fischer.)

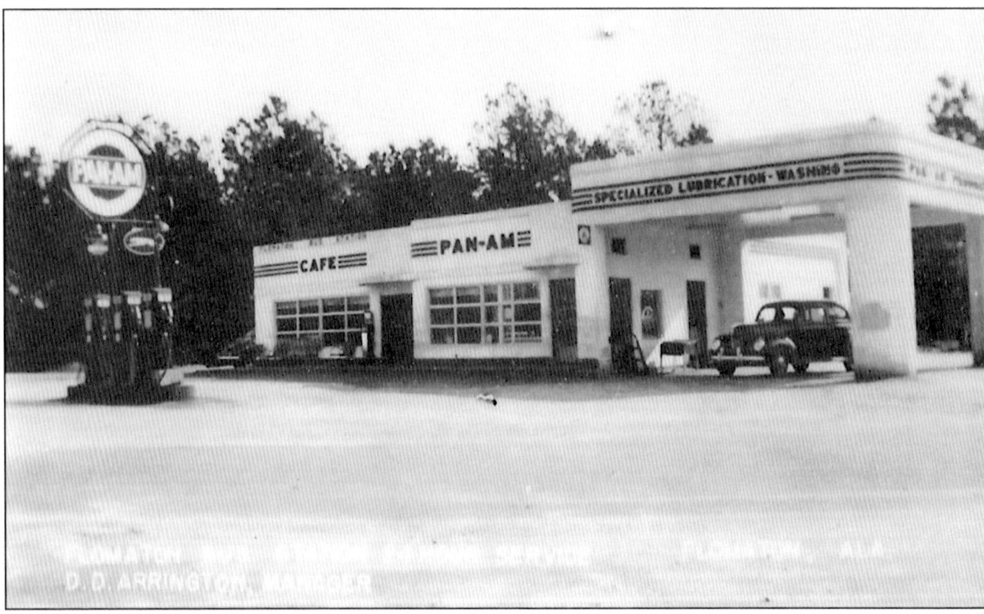

Arrington's Pan-Am was a service station and Greyhound bus stop on Highway 31 next to the bridge over Big Escambia Creek in the 1930s. It was owned by D.D. Arrington, who would become the mayor of Flomaton. Arrington's has been in business for over 100 years and is still in operation today in Flomaton. (Courtesy of Jerry Fischer.)

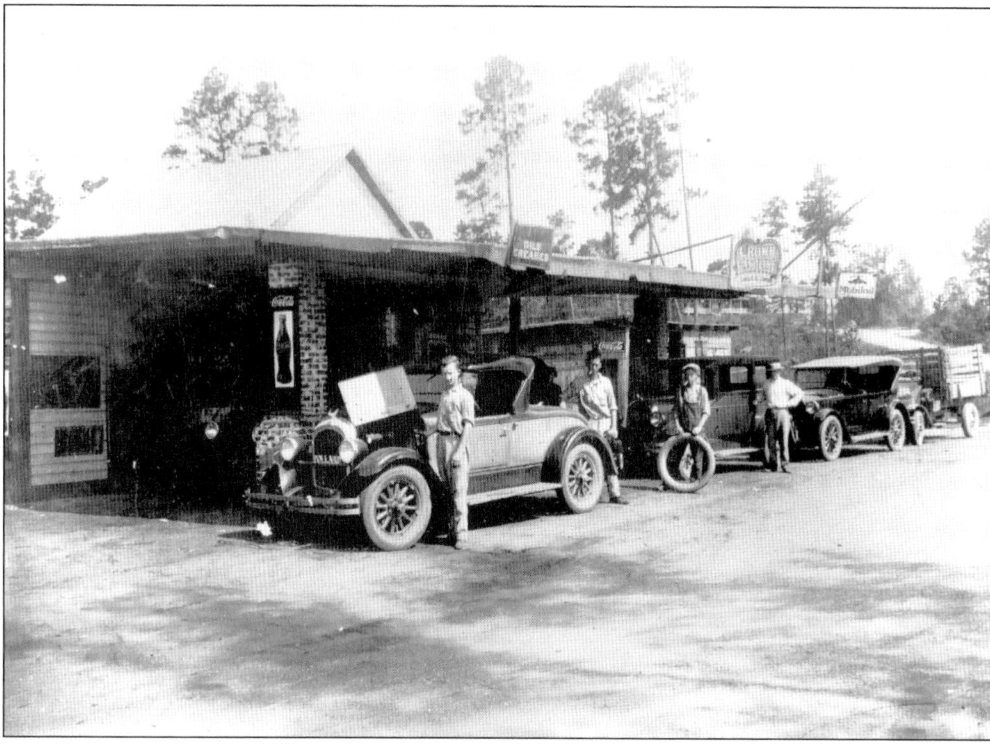

Congleton's Service Station was located on Highway 31 in Pollard. (Courtesy of Jerry Fischer.)

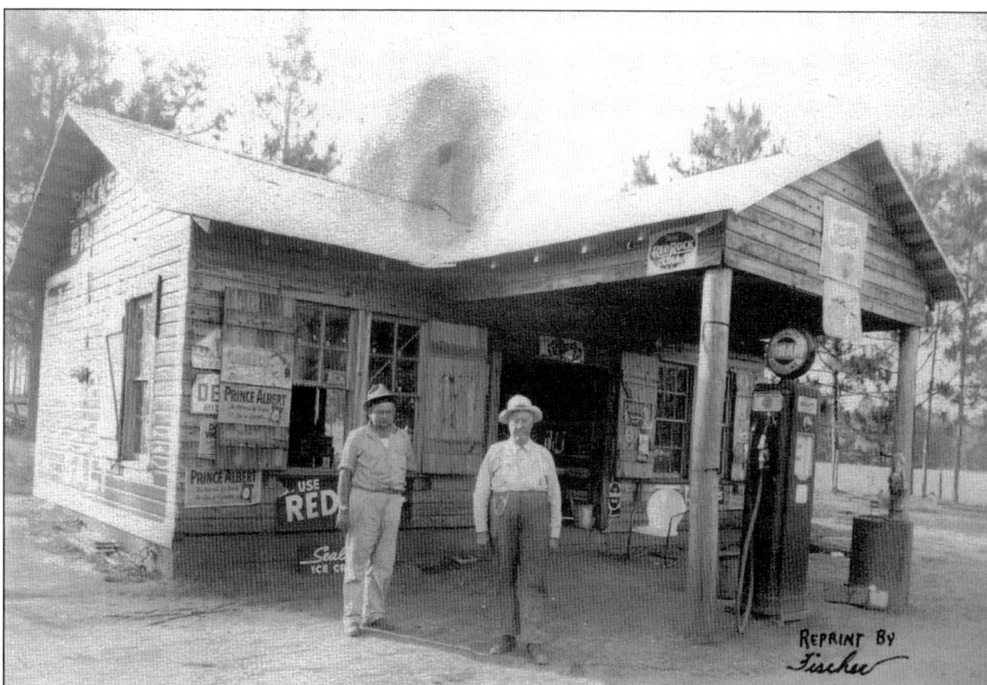

Woody Daw (left) and James Harrington stand outside of the McCall County Store in Flomaton. The store was located on Highway 31 before the road was paved. (Courtesy of Jerry Fischer.)

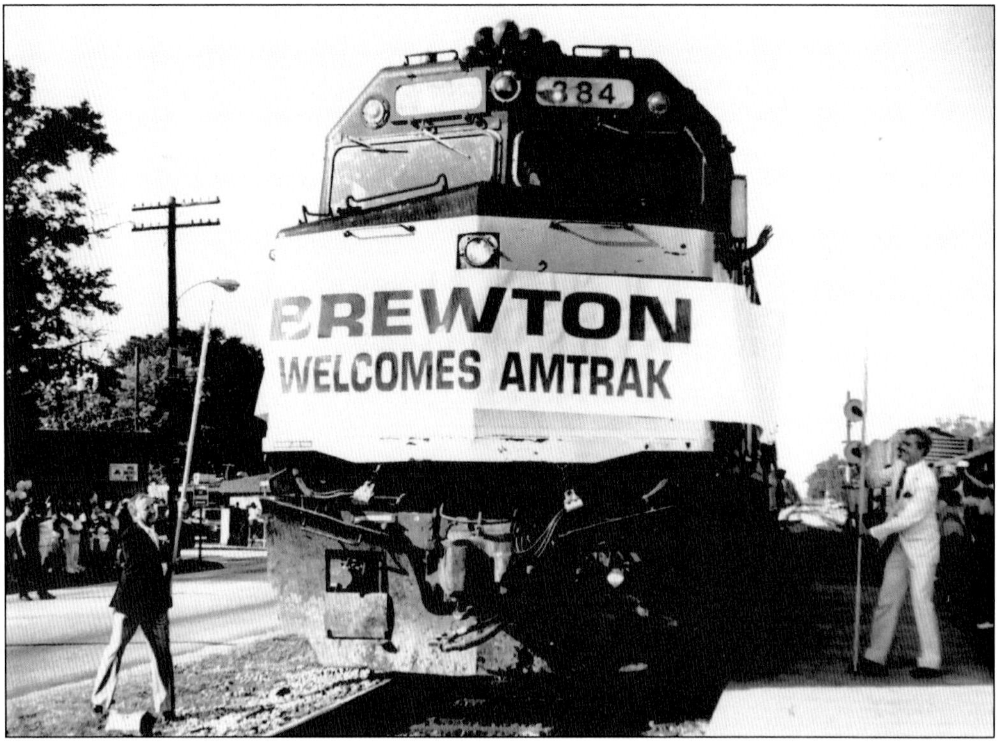

Mayor Ted Jennings (right) and Wick Leatherwood stand and watch at the AMTRAK inaugural stop in May 1991. The AMTRAK service was made possible by the hard work and determination of local attorney Jim Hart. (Courtesy of D.W. McMillan Trust.)

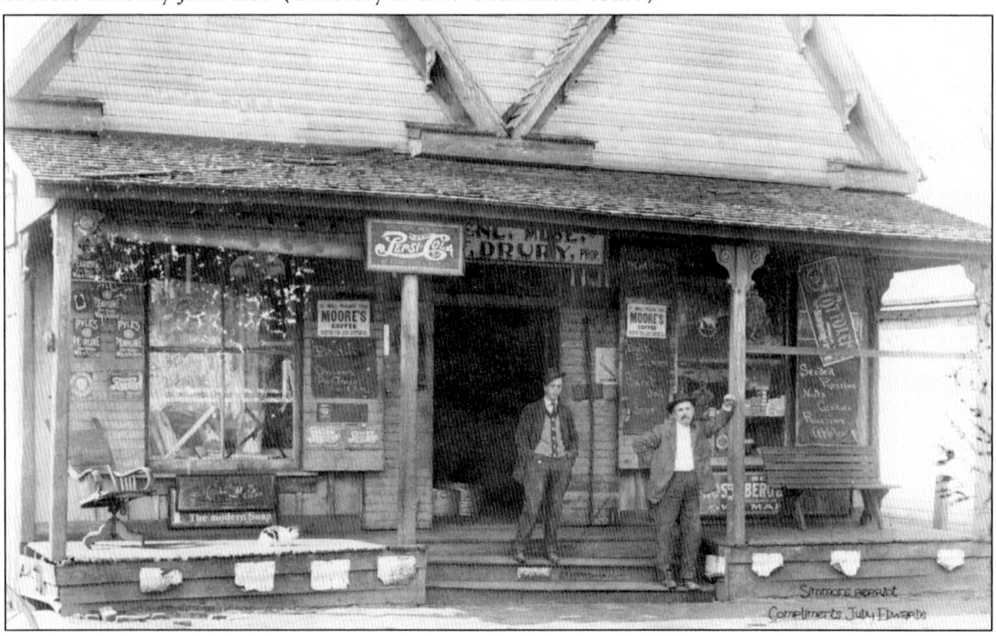

Dick Sowell Sr. (left) and John Francis Drury stand outside of the Drury Store in downtown Flomaton. Drury was a Civil War veteran and an early businessman in Flomaton. He closed his store for good after the 1929 flood. (Courtesy of Jerry Fischer.)

Six

Institutions and Organizations

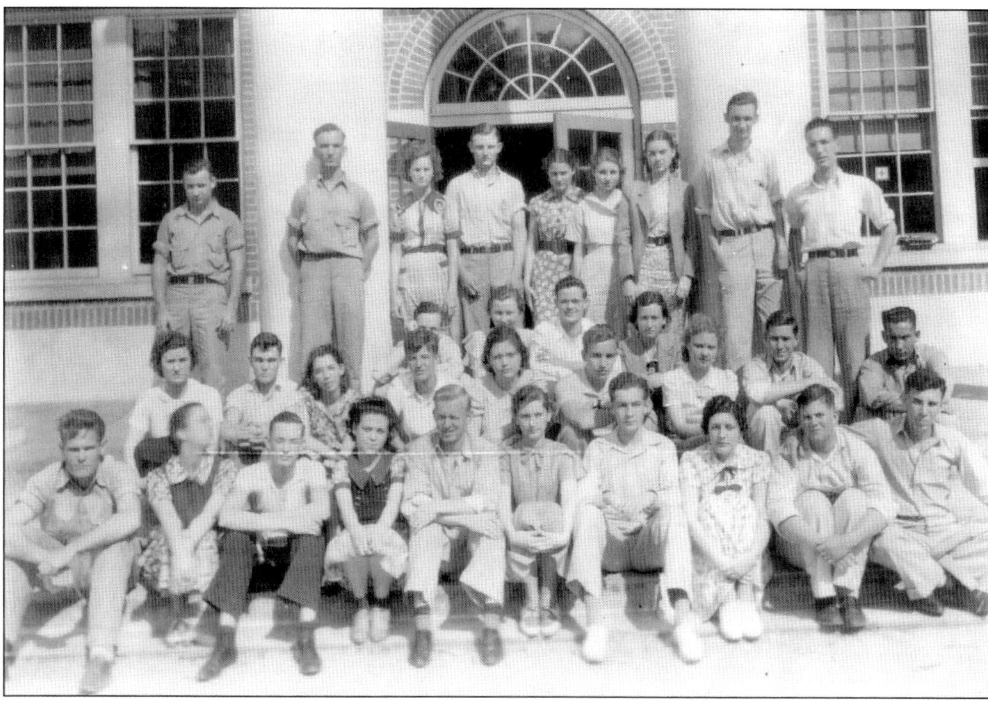

The Escambia County High School class of 1938 gathers in front of the oldest county high school in Alabama. Atmore had the first county high school in Alabama, which was built in 1903. From left to right are (first row) Ralph Stevens, Grace Boone, Kenneth Rollins, Agnes Martin, Francis Stone, Bessie Merriweather, Howard Patterson, Grace Freeman, Carol Crawford, and Russell Stewart; (second row) unidentified, James Robinson, Mary Jane Lloyd, Farrah Hall, Elsie Hardy, Braxton Bragg, unidentified, Roscoe Goldsmith, and Lindsey McGhee; (third row) John Lee Downing, unidentified, John Webb, and unidentified; (fourth row) Jessie Owens, unidentified, Helen Carnley, Johnny Moye, Bertha Hardy, Ida Hall, Grace Freeman, Ralph Patrick, and Tony Albert. (Courtesy of John Webb.)

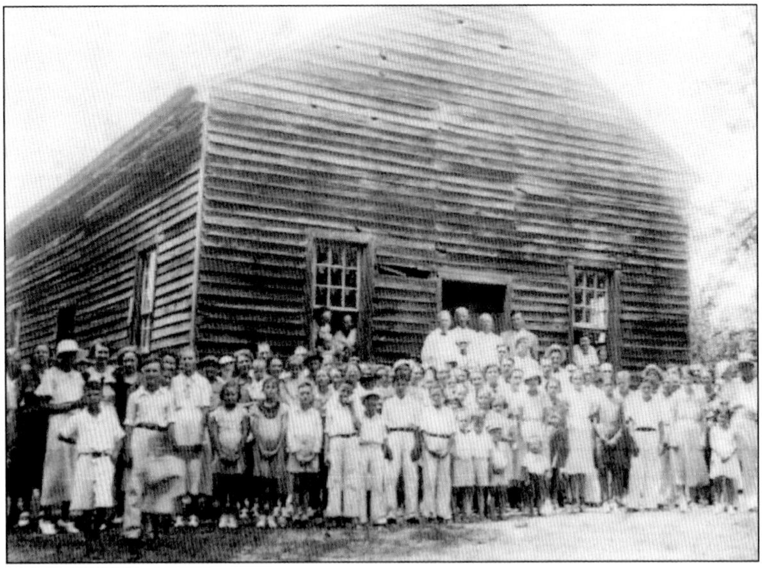

This First Baptist Church on Cowpen Creek Road in Canoe appears during the 1890s. It was one of the original churches in Canoe and was located north of the railroad near what is now Escambia Academy. (Courtesy of Carolyn Conn and Kevin McKinnley.)

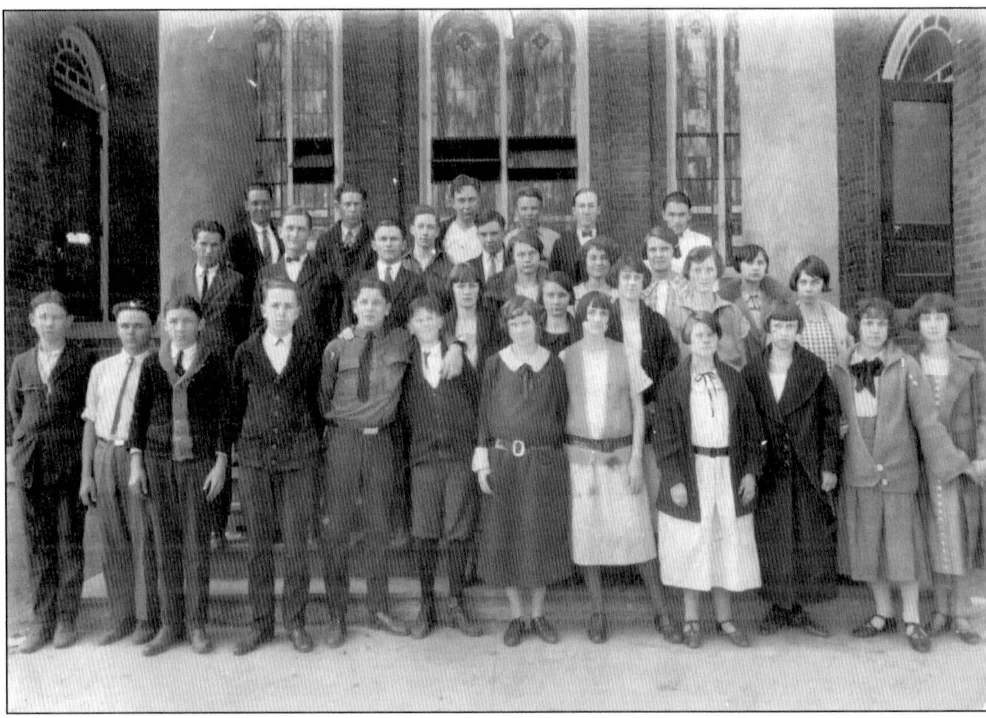

This is the Escambia County High School class of 1927. The group is assembled in front of the First Baptist Church, one of many area churches that doubled as classrooms after the Atmore School burned in 1925. Lillian Porter (second row, fourth from right) was such a renowned teacher of math and science that the school attracted students from other counties. Her reputation was such that she was recruited by the US government and sent to Oak Ridge, Tennessee, to work on the Manhattan Project, which developed the atomic bomb. (Courtesy of John Webb.)

The US Army's Rainbow Division marches through Canoe in 1917 (during World War I) near Lowrey Crossing, marked by the railroad crossing sign in the background. The regiment was under the direction of Col. Douglas MacArthur. The spot was also the site of camps for both the North and South during the Civil War. On their way from Pensacola to Blakely in Baldwin County, 8,000 Union soldiers spent the night in a torrential downpour. When they could find nothing but wet wood, they broke open cartridges and used gunpowder to start a fire. (Courtesy of Kevin McKinnley.)

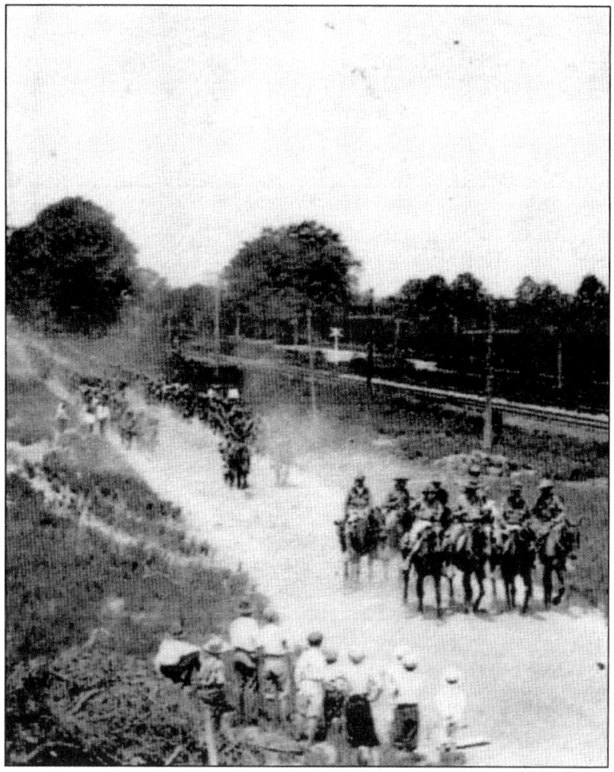

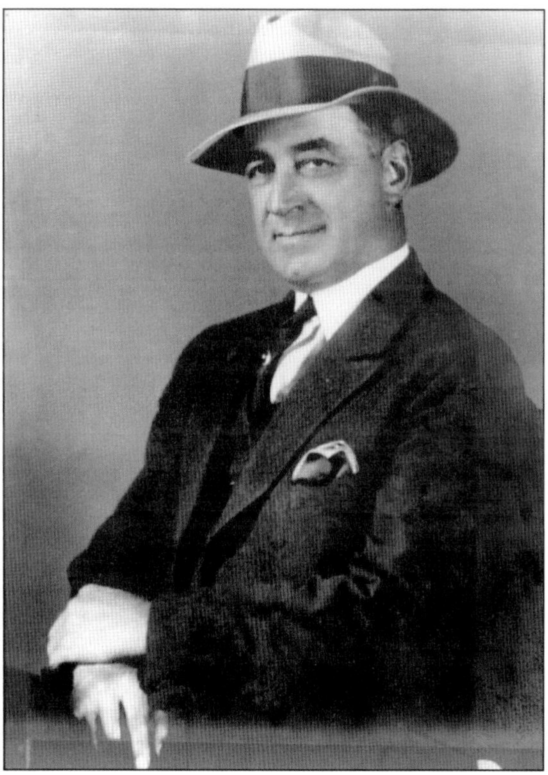

Dr. D.W. McMillan was born in Escambia County, educated in the United States and Europe, and served as a major in the Medical Corps of both Great Britain and the United States during World War I. He practiced medicine in Pensacola, Florida, and left his estate to the poor and needy of both Escambia County, Alabama, and Florida. The hospital in Brewton is named in his honor. (Courtesy of D.W. McMillan Trust.)

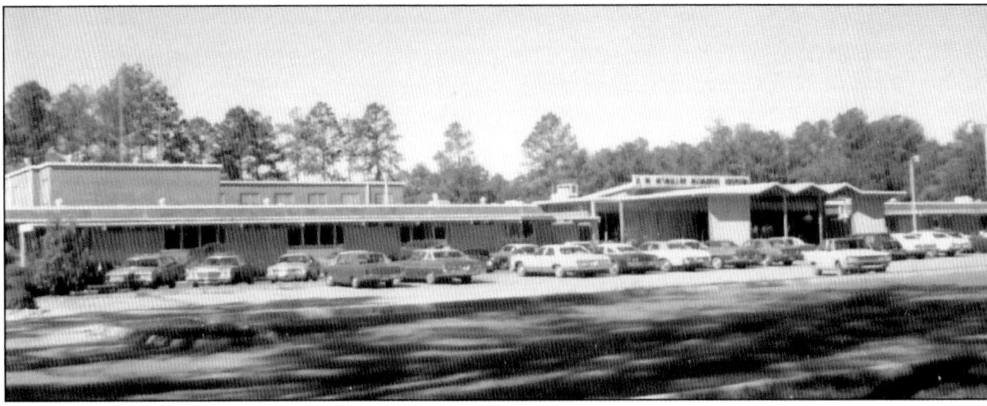

This is the D.W. McMillan Hospital. It opened in 1953 with 35 beds, operating rooms, and an emergency room. It has been greatly expanded since those early days and had 91 beds at one time. (Courtesy of D.W. McMillan Hospital.)

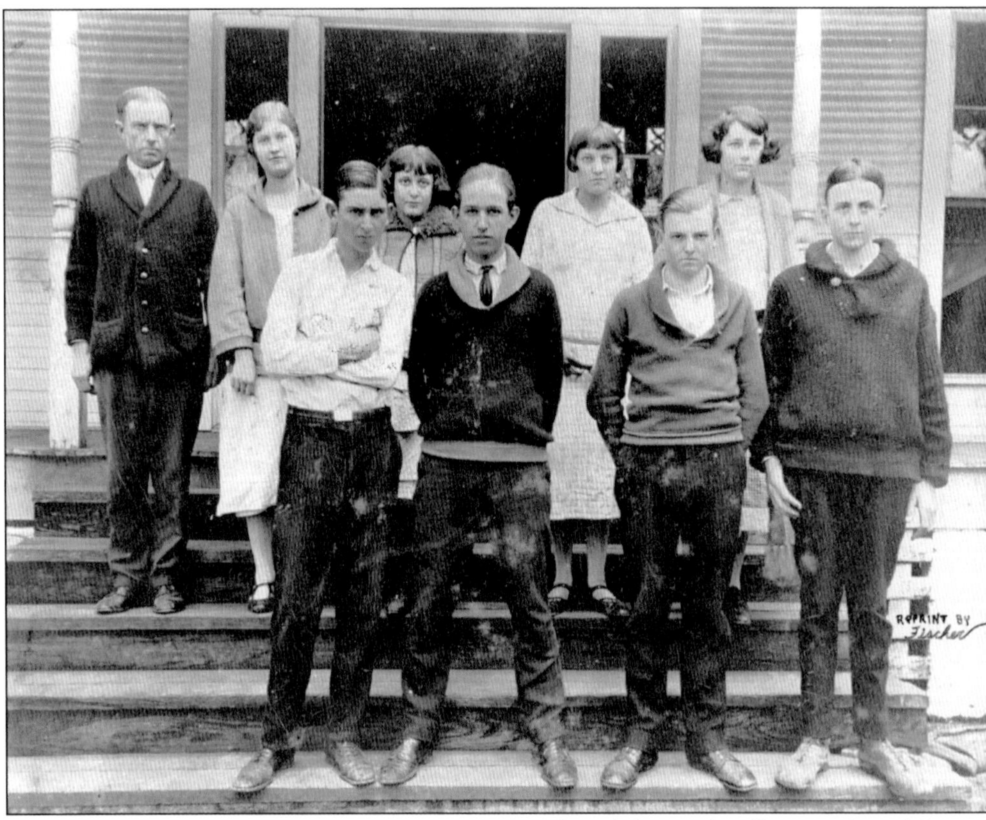

The Pollard School senior class of 1925 consisted of, from left to right, (front row) Paul Owens, Walter D. McClellan, James "Jim" Wyche, and Henry Edwards; (back row) principal Drury Leroy Walker, Callie Harris Wilson, Birdie Fuller, Lucile Walton, and Jessie Lee Driskell. (Courtesy of Jerry Fischer.)

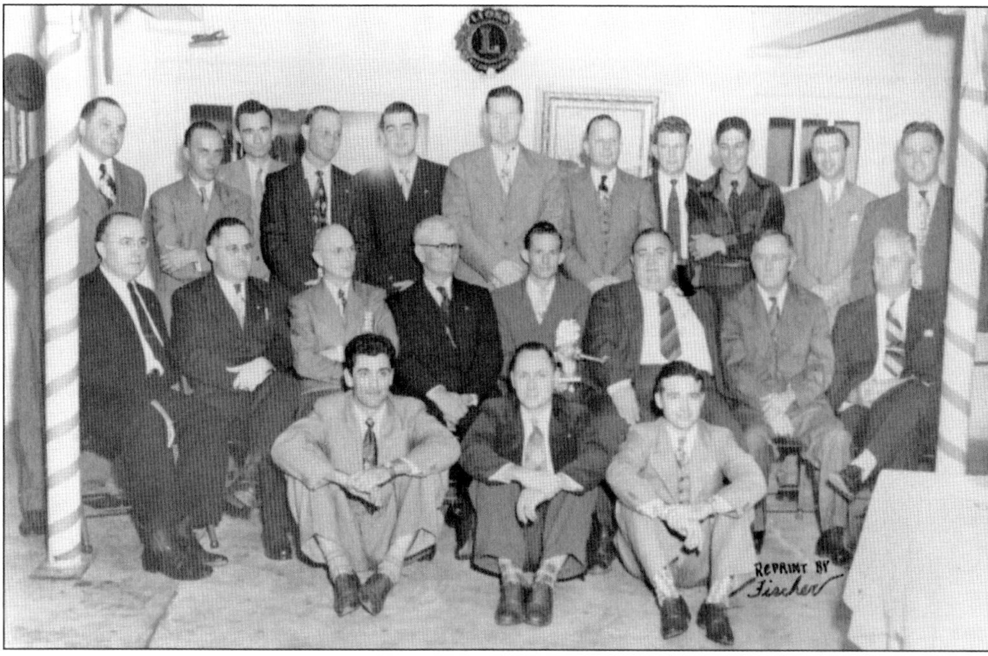

The Lions Club at Flomaton was the biggest club in town and was a huge supporter of the community. Pictured are various community leaders who made the Flomaton Lions Club so successful. (Courtesy of Jerry Fischer.)

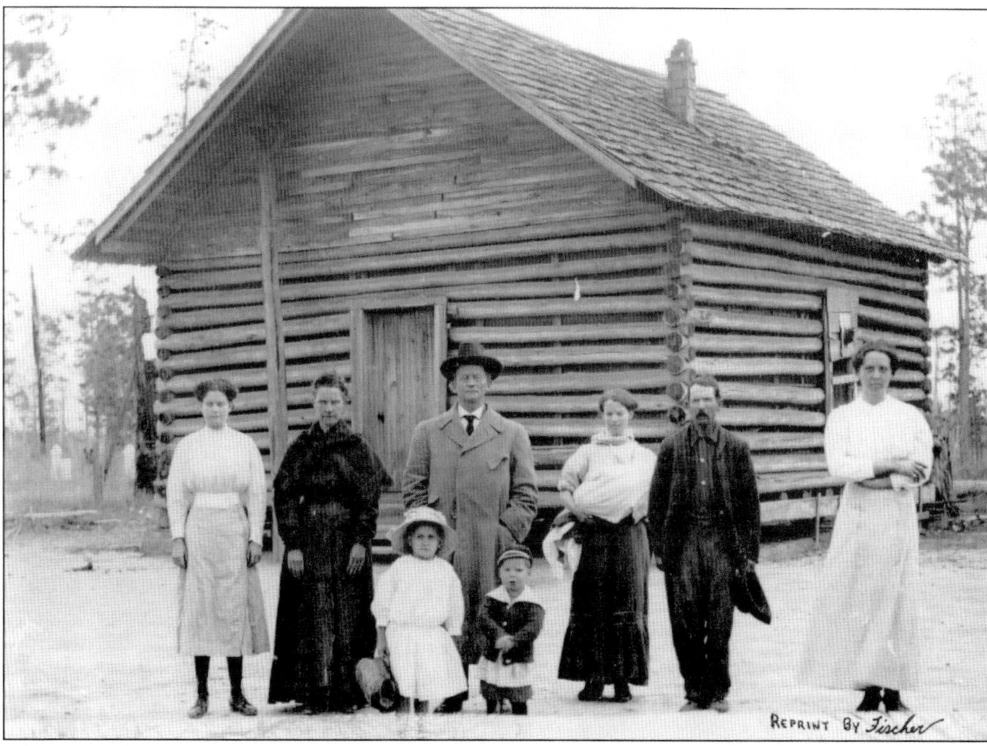

The Mize family stands outside the old church in the Pineview community. (Courtesy of Jerry Fischer.)

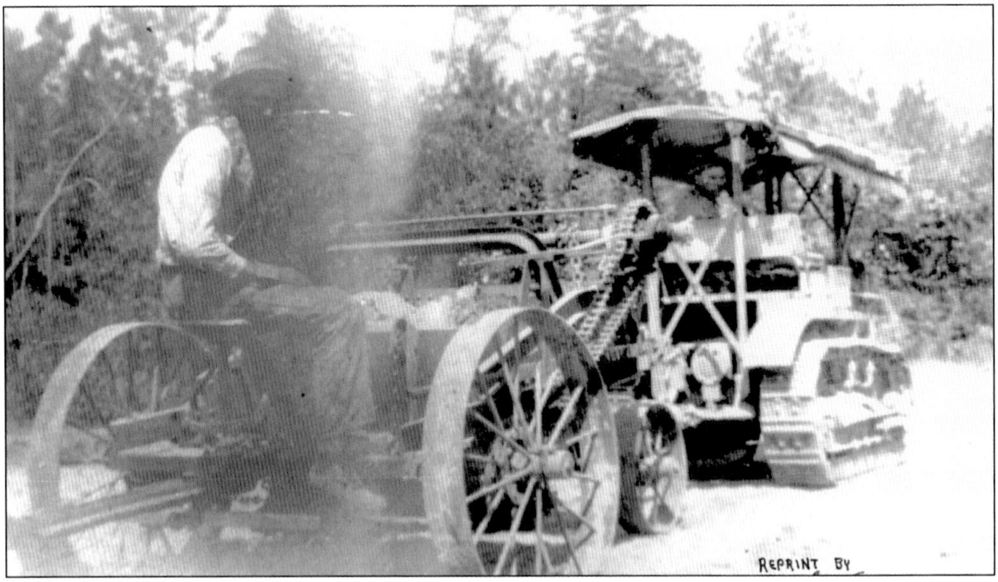

This picture was taken in February 1928 and given to Devon Wiggins, chairman of the Escambia County Commission, on May 19, 1983, by C.E. (Charlie) Dawson, who is a retired employee of the road department in Escambia County, Alabama. The Caterpillar tractor and road machine were bought in December 1927. The road-machine operator is C.E. Dawson, and the Caterpillar tractor operator is Alton Godwin. At this time, there were three men working in the District No. 3 Road Department: the two mentioned here and Bruce Gatewood. There were no paved roads in Escambia County at that time. (Courtesy of Jerry Fischer.)

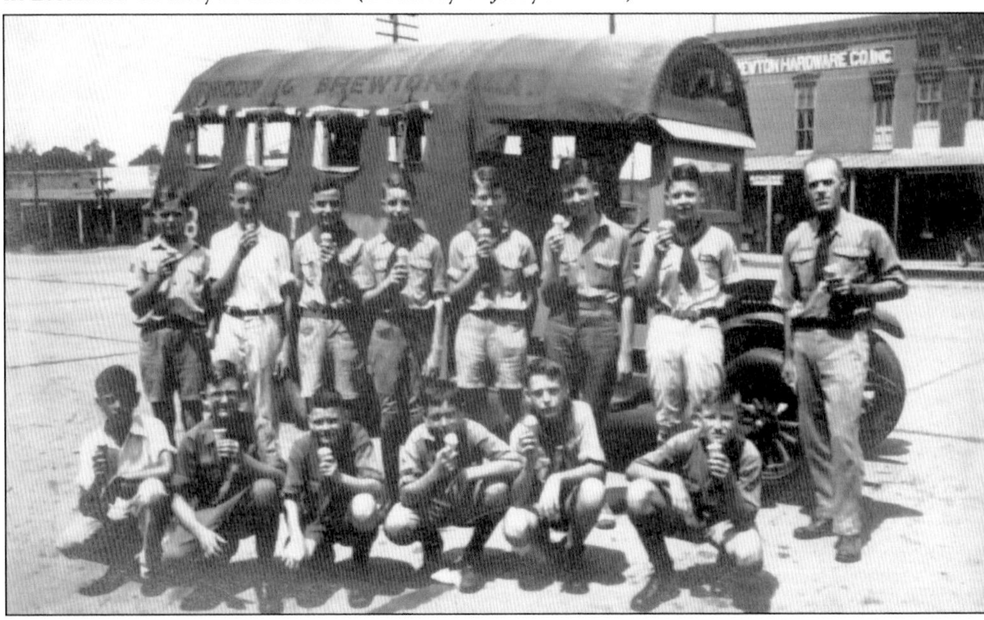

Boy Scouts from Troop 16 of Brewton, Alabama, are enjoying ice cream and posing for a picture in 1929. From left to right are (first row) Joe B. Luttrell, Brooks Hayes, John Douglas, Ben Strain, Pat Sowell, and Tommy Liles; (second row) Junior Mayo, Trav Hagood, O.C. Weaver Jr., Cleve Pittman, Clifton Jordan, Rufus Nettles, Billy Douglas, and C.P. Wrigley (preacher of the Presbyterian Church). (Courtesy of Buddy Liles.)

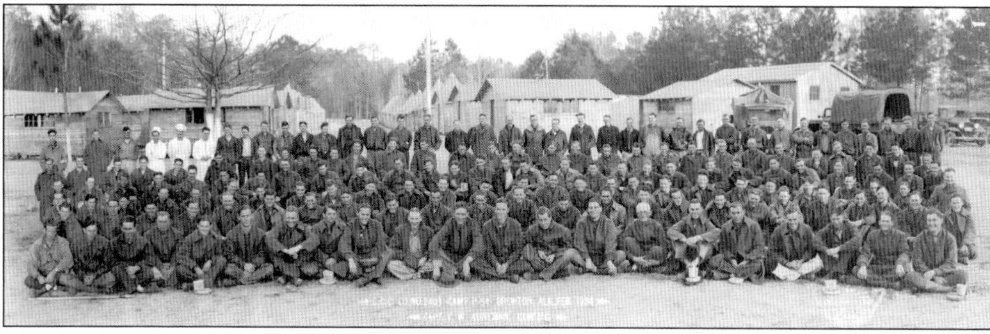

The Civilian Conservation Corps (CCC) camp in Brewton was located along the west side of Burnt Corn Creek near the present-day Arby's. Most of the work in this area was performed in the Conecuh National Forest in the eastern part of Escambia. The CCC improved roads, dug firebreaks, planted trees, and generally improved the forest. (Courtesy of Escambia County Historical Society.)

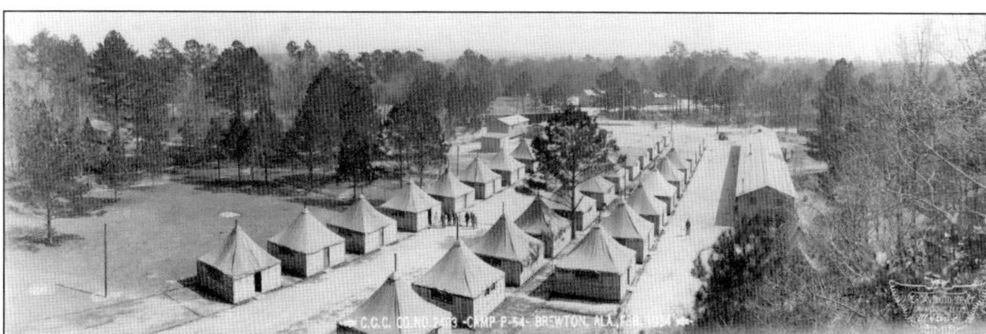

Shown is the CCC Company No. 2403's Camp P-54 in Brewton, Alabama, in February 1934. The CCC built up the nation's infrastructure and provided work for people during the Depression. Men were paid $30 per month plus room and board. (Courtesy of Escambia County Historical Society.)

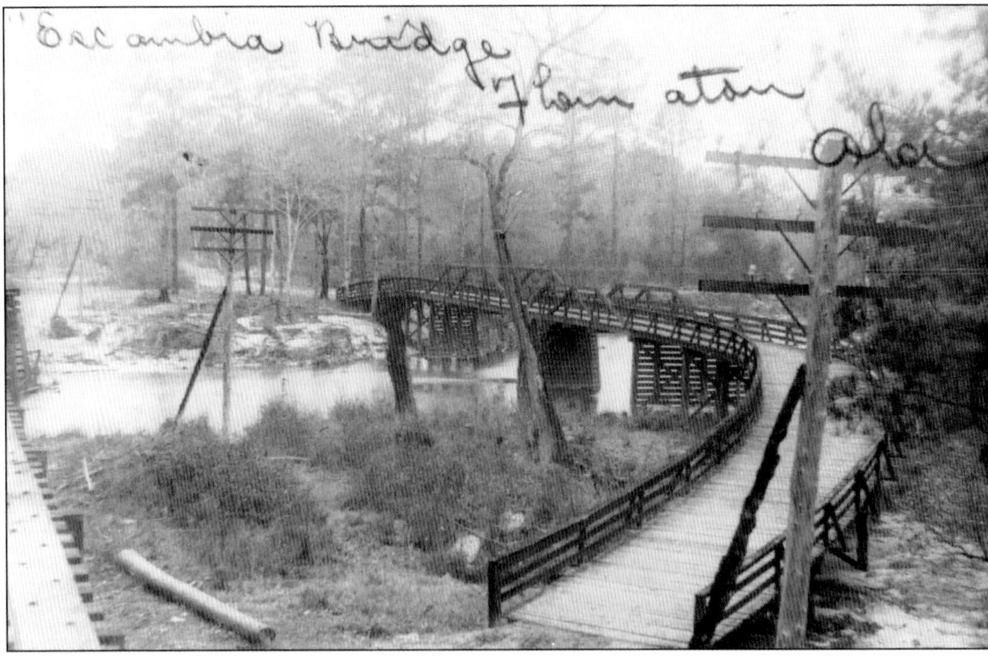

This image is looking west across the bridge at Big Escambia Creek on the road from Flomaton to Fannie. The road went under the railroad tracks, seen here in the bottom left corner of the picture. (Courtesy of Jerry Fischer.)

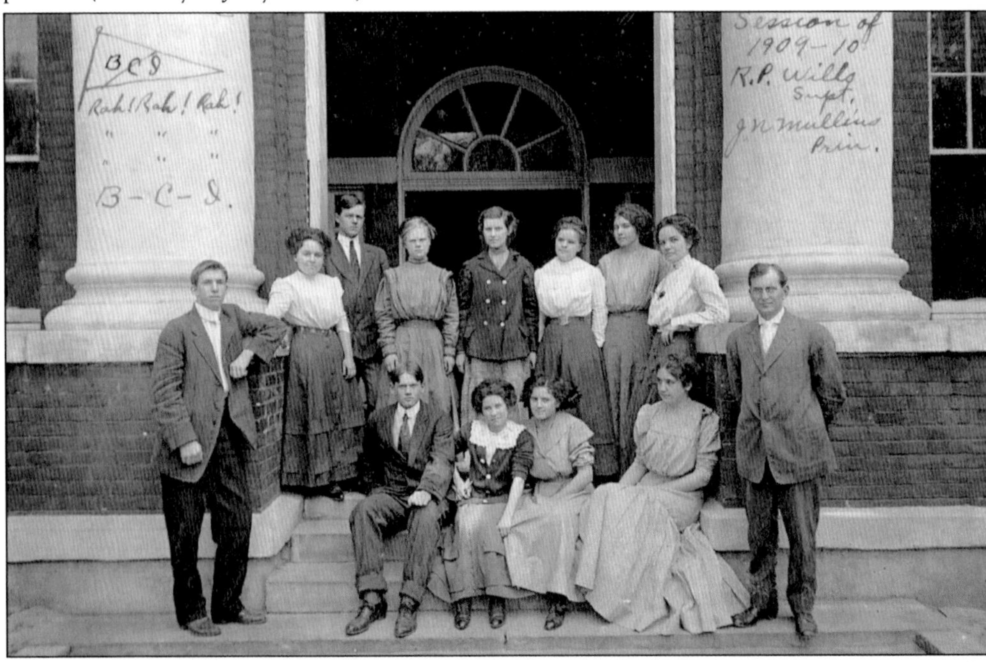

The Brewton Collegiate Institute senior class of 1909–1910 poses at the front entrance to the school; J.R. Mullins was principal. From left to right are (first row) Reeder Morris, Chao Robbins, Anna S. Downing, Elanor Davison, Rufus Mayo, and R.P. Willie; (second row) Mimie M. Harold, Joe Sowell, Melba Robbins, Mary Stewart, Bernice Larkin, Bessie Wiggins, and Annie Corrie Park. (Courtesy of Escambia County Historical Society.)

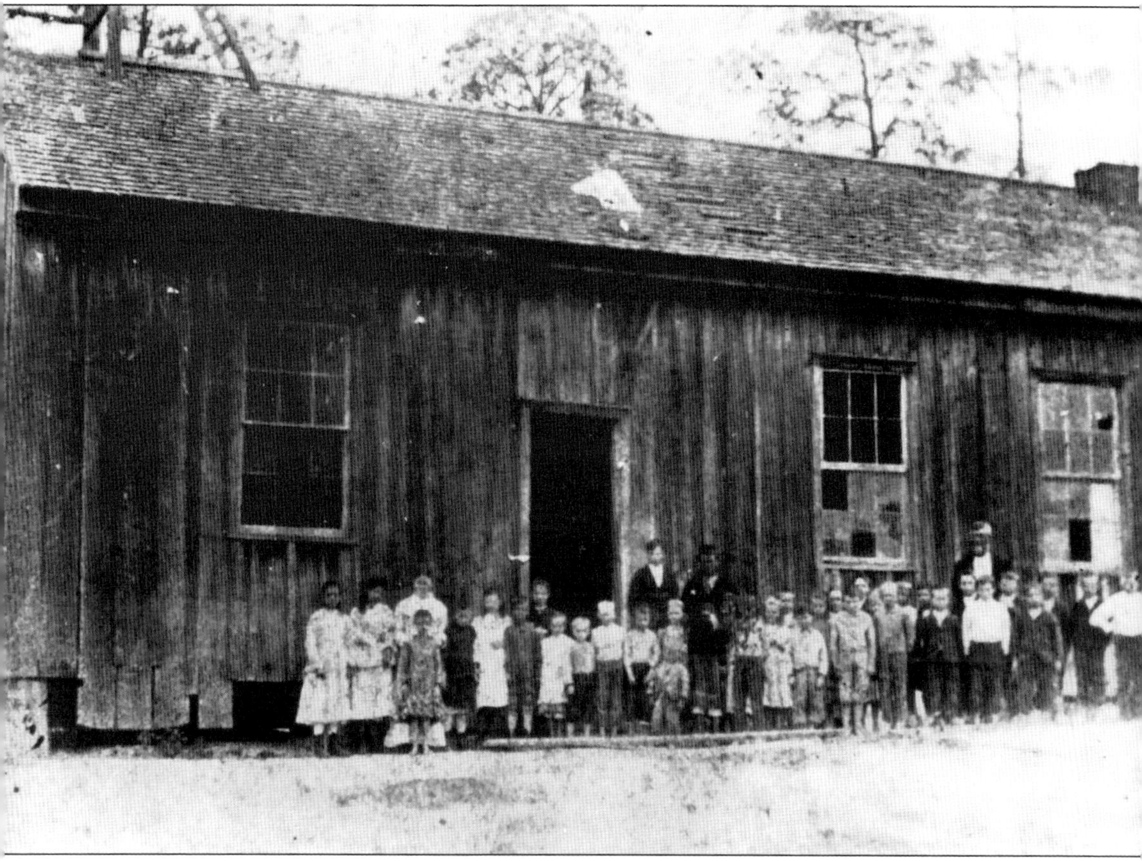

Old Brewton Academy was one of the first schools in Brewton. It was located where T.R. Miller Mill Company sits today, probably near the depot. (Courtesy of Escambia County Historical Society.)

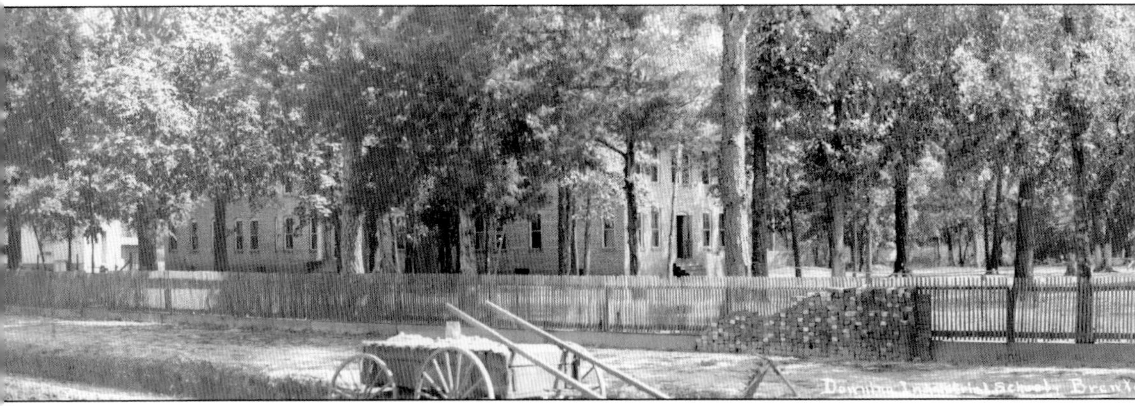

The Downing Shofner Institute in East Brewton was a place of beauty in its day. (Courtesy of

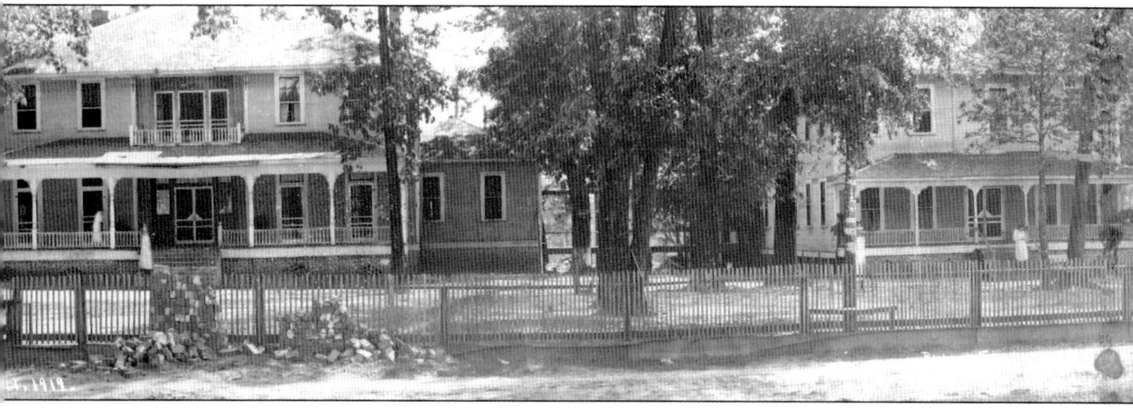

Escambia County Historical Society.)

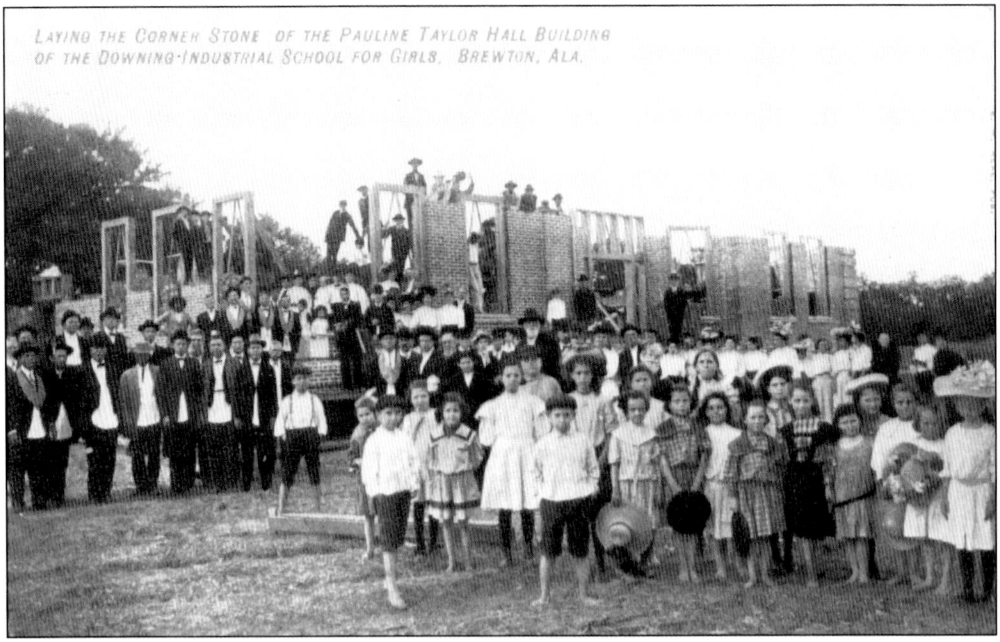

The school has been known by many names, including the Downing Industrial School for Girls. It was an instant success and continued operations from 1904 to 1943. (Courtesy of Yank Lovelace.)

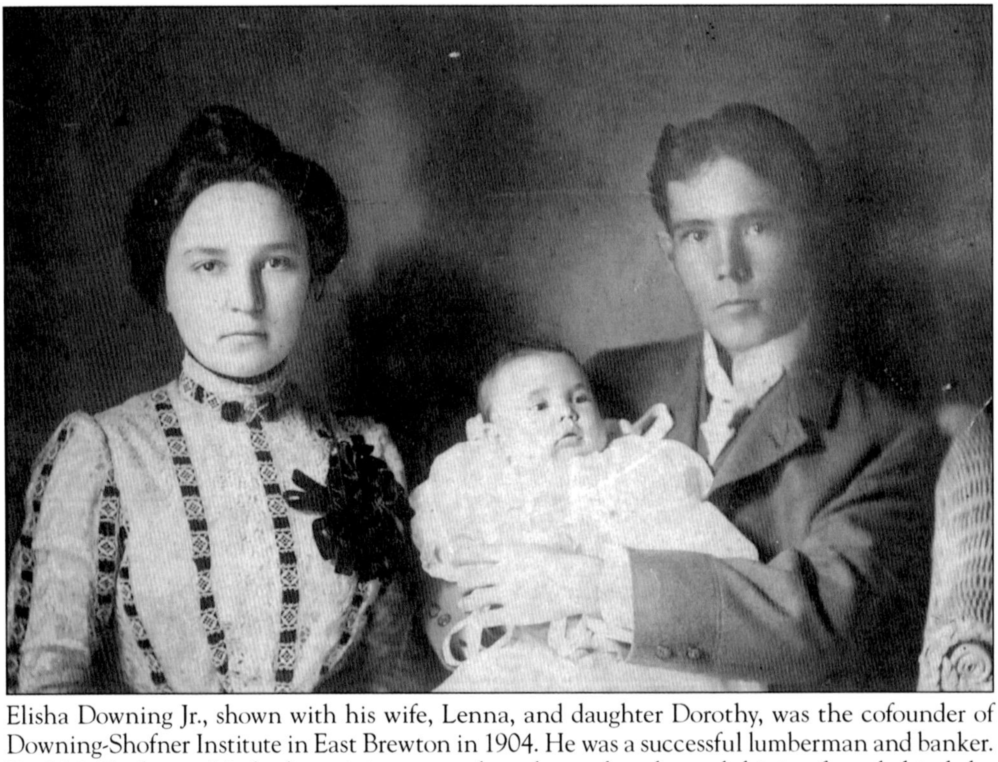

Elisha Downing Jr., shown with his wife, Lenna, and daughter Dorothy, was the cofounder of Downing-Shofner Institute in East Brewton in 1904. He was a successful lumberman and banker. Dr. J.M. Shofner, a Methodist minister, was the other cofounder and driving force behind the girl's school. (Courtesy of Escambia County Historical Society.)

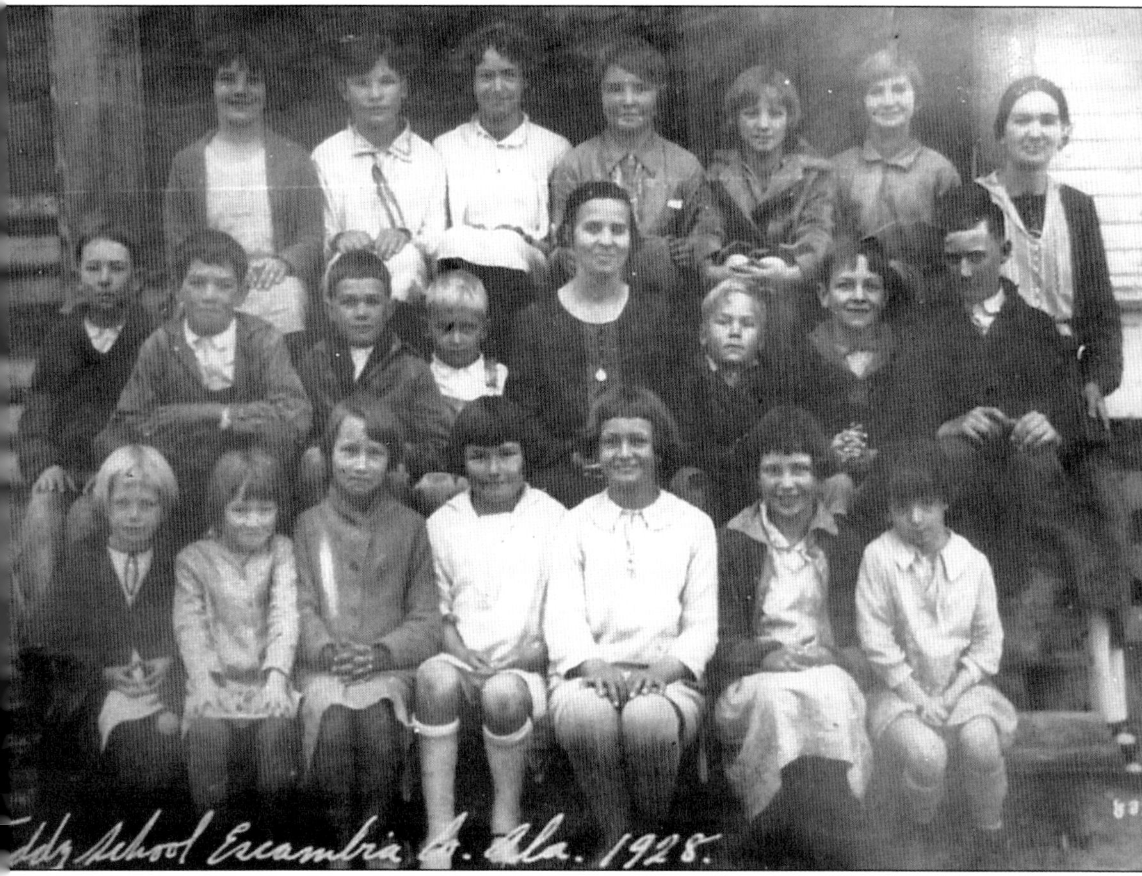

Teddy School was located in northeast Escambia County near the Conecuh River. This group of students of all ages represented the entire student body in 1928. Most of the county schools no longer exist. They have been consolidated, and the small towns are now mainly bedroom communities. (Courtesy of Escambia County Historical Society.)

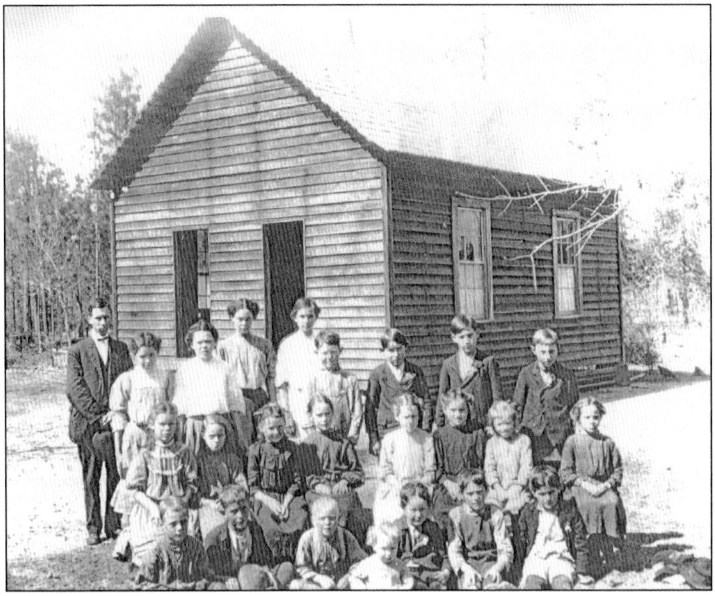

Schoolchildren are standing outside the county school building near Barnett Crossroads for their school picture. (Courtesy of Jerry Fischer.)

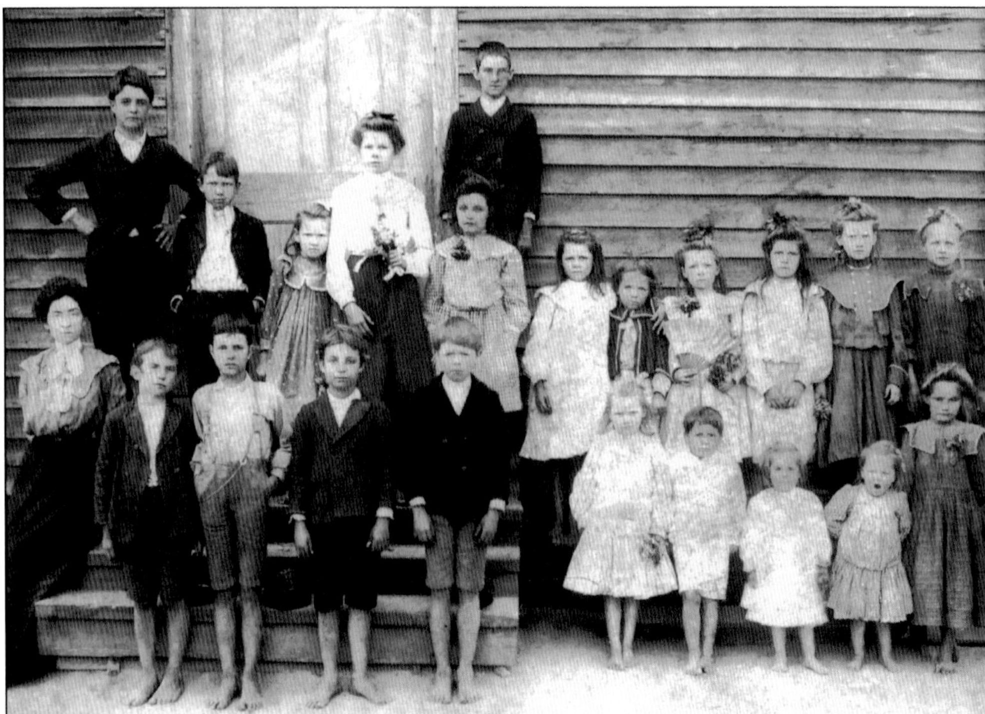

The children might be barefooted, but they are well dressed and groomed for this photograph taken in 1905. These Wallace School students include, from left to right, (first row) teacher Carrie McPhail, Clyde Owens, Joe McDavid, Mallory McDavid, Drew Elder, Viola Elder, Eugene Hines, Bertha Hines, Agnes Elder, and Lena Cardwell; (second row) Wheeler McDavid, Lee Elder, Willie Elder, Gertrude Hines (holding flowers), Clyde Webb (leaning against back wall), Minerva McDavid (in front of Clyde), MaeBelle Hines, Mammie Elder, Lucy Elder, Esther Hines, Rowena Elder, and Dora Cardwell. (Courtesy of Anne Torreins and Escambia County Historical Society.)

After losing all four of his brothers in the Civil War, W.S. Neal moved to Brewton in 1881 and served as a teacher and then 23 years as superintendent of education. The W.S. Neal School in East Brewton is named for him. He sponsored legislation that resulted in the first county high school in Alabama being established in 1903. (Courtesy of Escambia County Historical Society.)

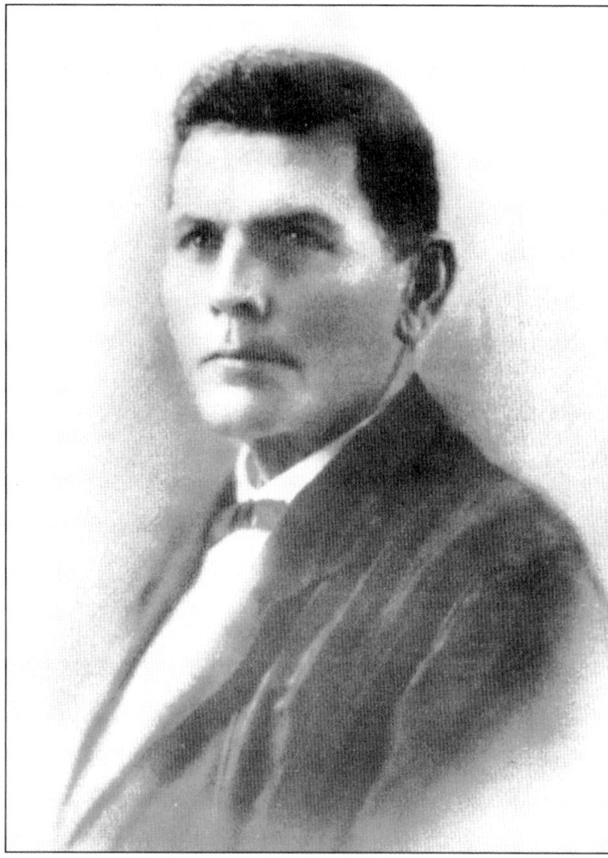

W.T. Neal Sr. is dedicating the library and presenting a picture of his father, W.S. Neal, at the W.S. Neal High School in East Brewton. O.C. Weaver, county superintendent at the time, is sitting just to the left of the portrait. (Courtesy of Escambia County Historical Society.)

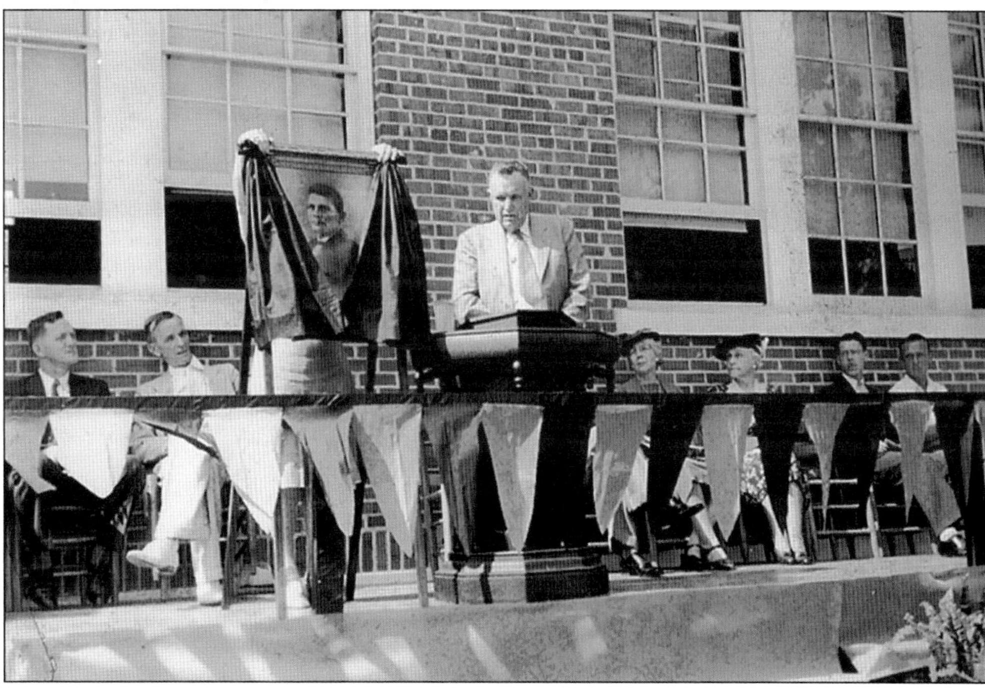

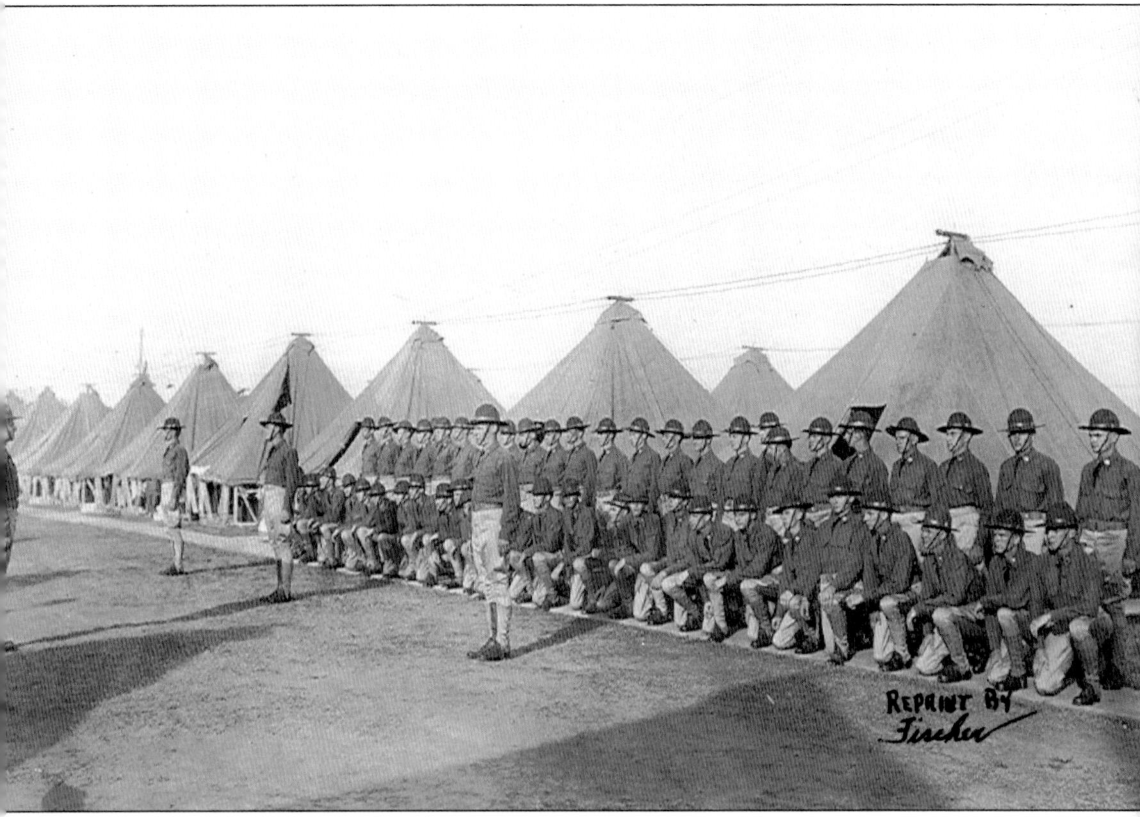

Company G of the Alabama National Guard camped near downtown Brewton in 1917. The soldiers were on their way to Montgomery to enter the war. (Courtesy of Jerry Fischer.)

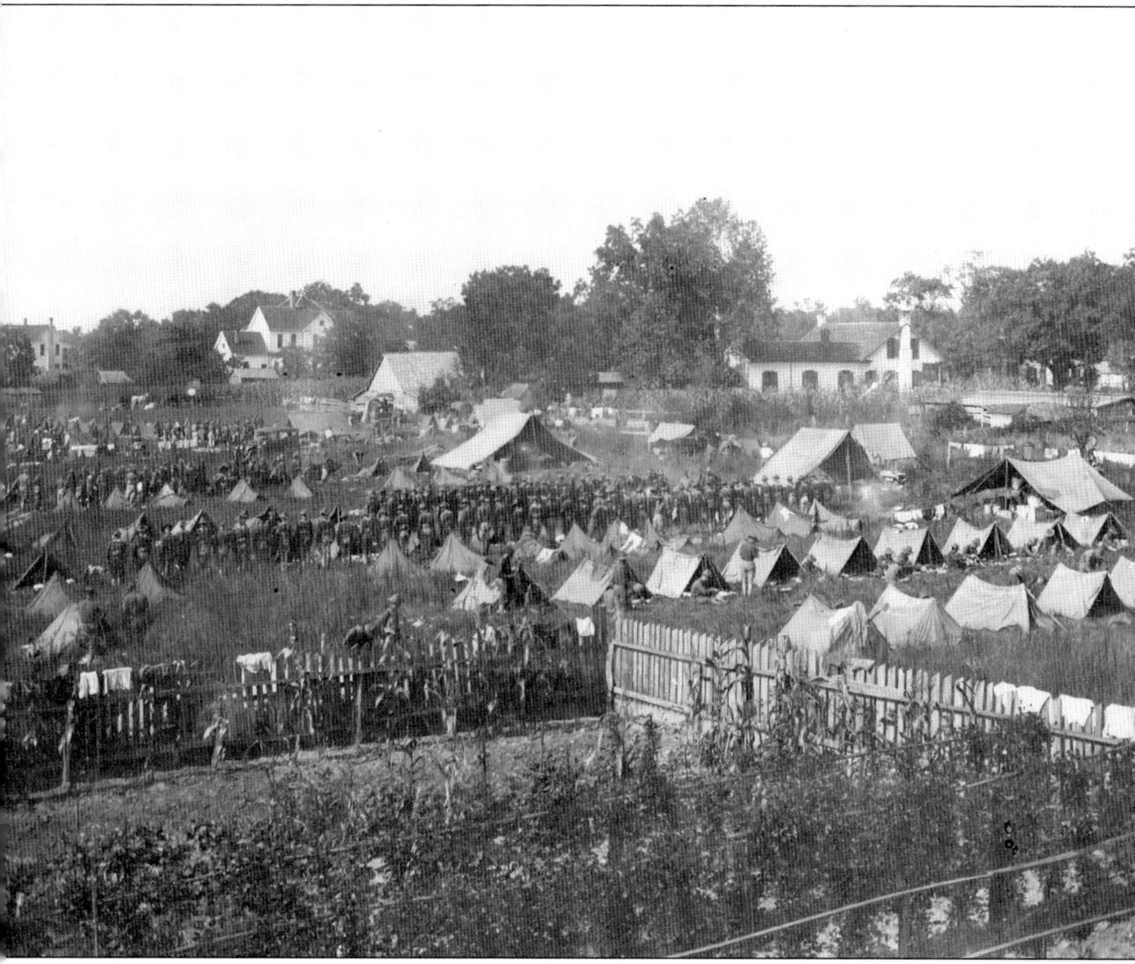

Company G of the Alabama National Guard in Brewton is seen on the way to Montgomery for deployment to Europe. The United States of America declared war on Germany in 1917; this image was taken that year. The camp was on the east side of US Highway 31 near present-day Granberry Street. The three large homes in the background are, from left to right, the W. S. Neal home, the Alto Vela Lovelace home (later to become Dr. Al Holley's hospital), and the Sowell home. (Courtesy of D.W. McMillan Trust.)

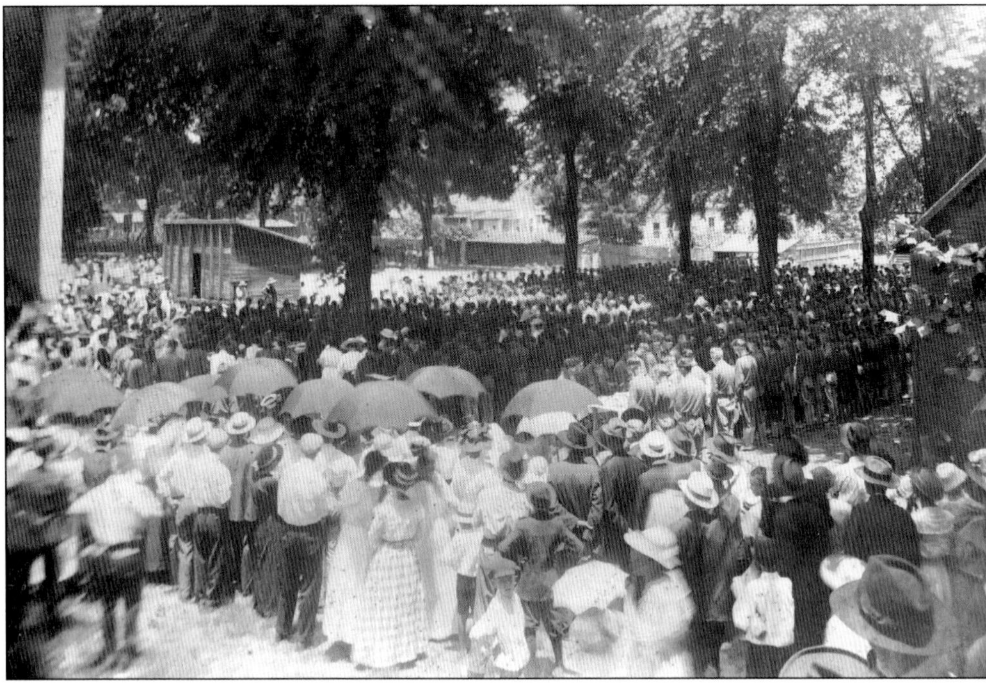

When Company G of the Alabama National Guard moved through Brewton in 1917 on the way to be sent to war, the soldiers were given a picnic dinner by residents on the school grounds next to the courthouse. (Courtesy of D.W. McMillan Trust.)

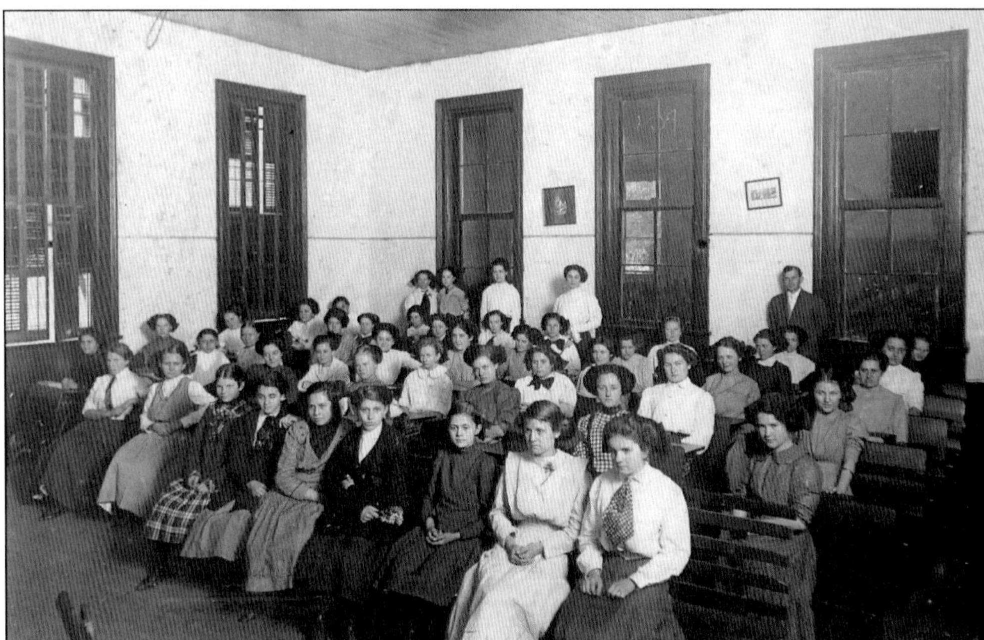

The Brewton Collegiate Institute opened in 1886. It was a private school established by a group of residents interested in education. It operated until 1925, when public schools began to offer a good education. These 1909–1910 high school girls are shown in one of the classrooms. (Courtesy of Lydia Grimes.)

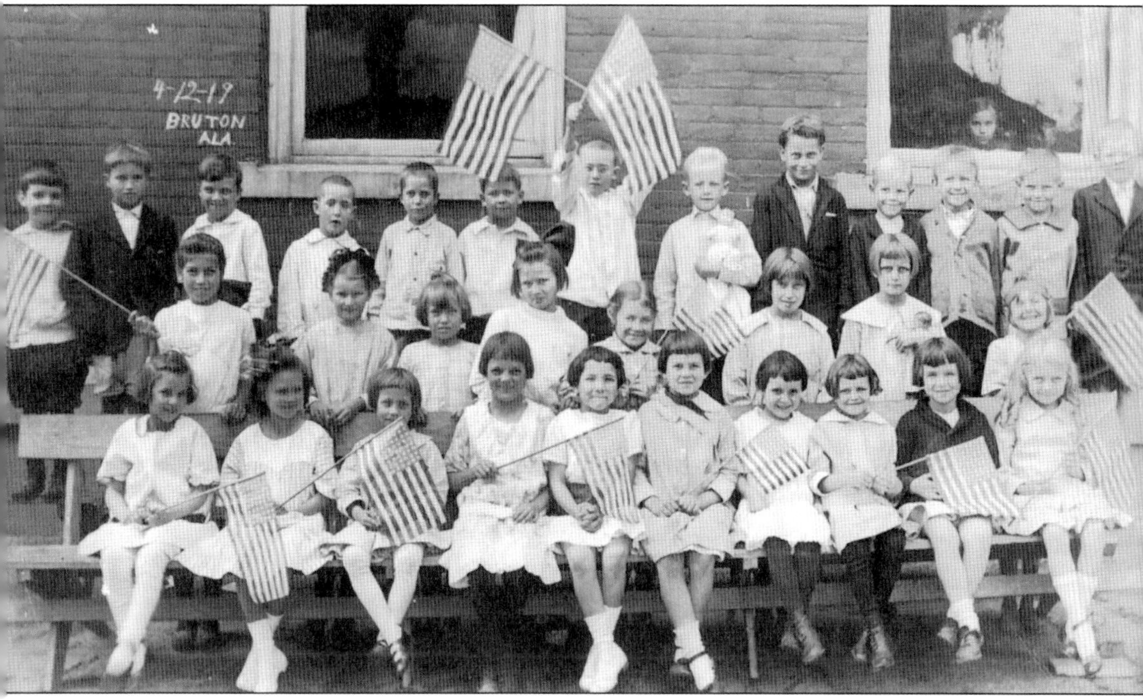

This photograph of a patriotic group of happy schoolchildren was taken in 1919. The spelling of the town's name came from the Brewton family. At one time, the family spelled its name "Bruton." (Courtesy of Escambia County Historical Society.)

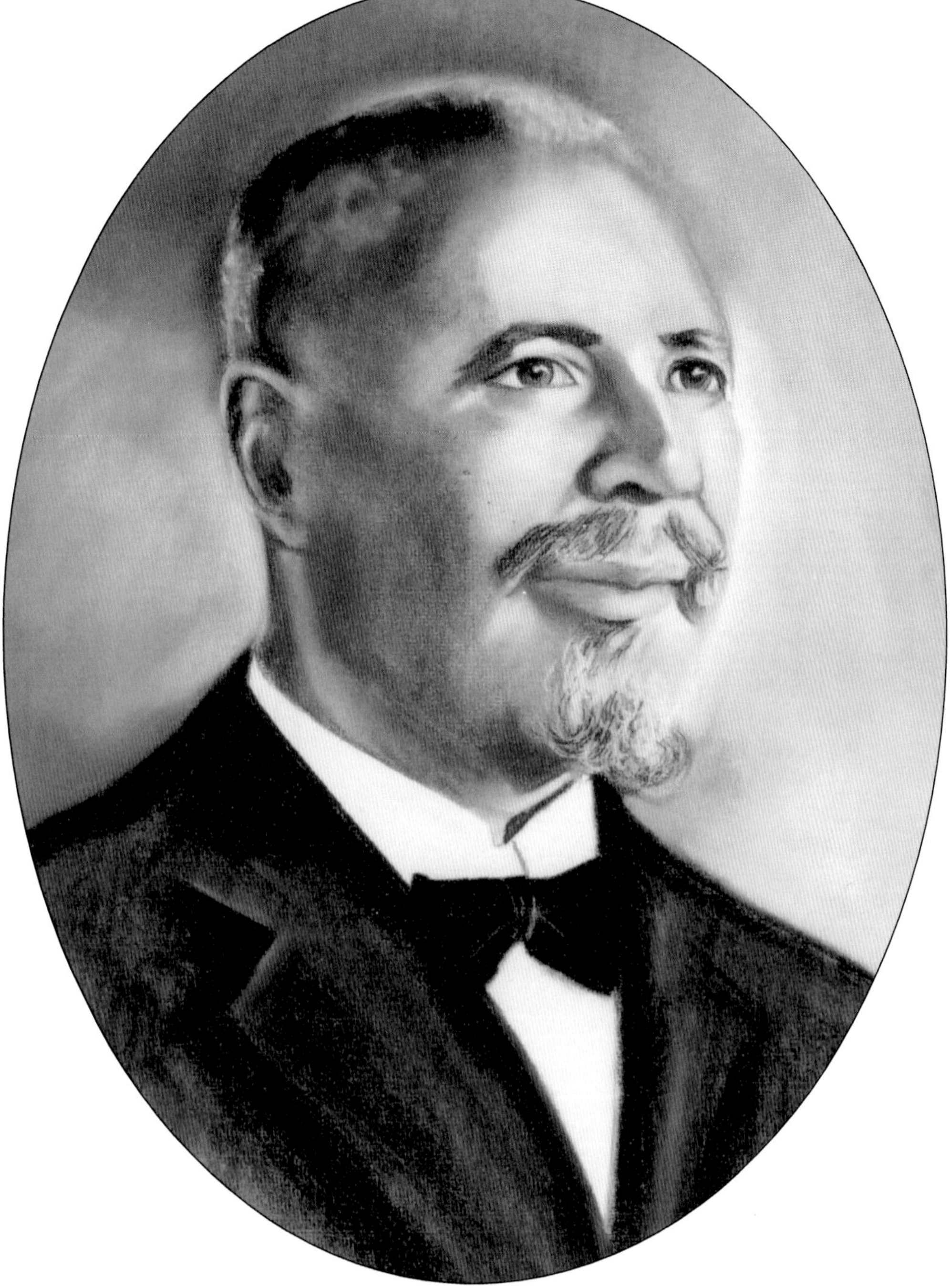

Dr. James Dooley Sr., an educator from Georgia, founded Southern Normal School in Brewton in 1911. He wanted to provide a solid educational program with a Christian foundation to the African American community. The Reformed Church in America helped to support the school as a mission station in 1919 and continued that support for many years. The school has struggled in recent times but is now part of Alabama State University with a new mission and new vision for the future. (Courtesy of Escambia County Historical Society.)

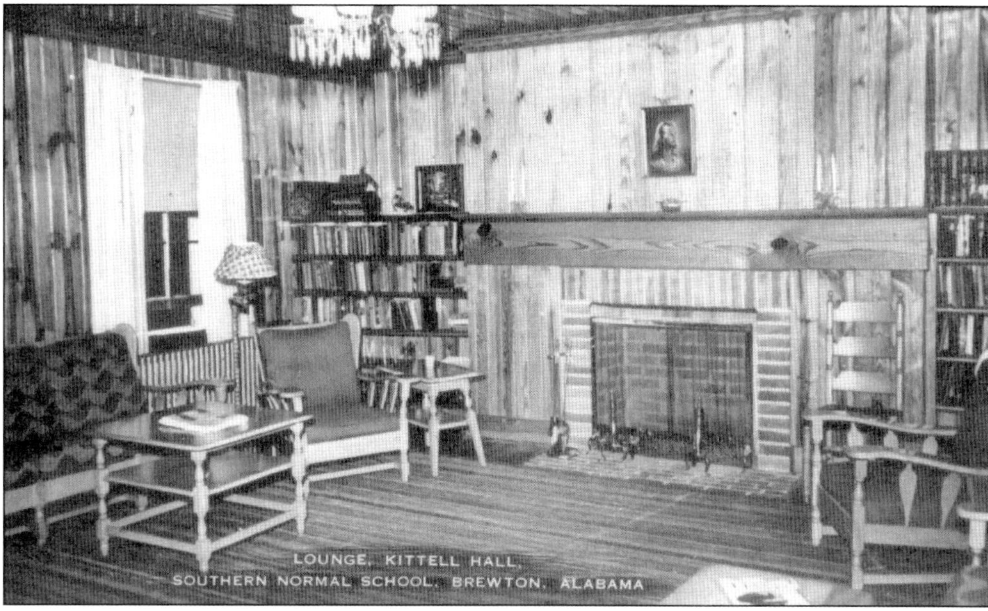

Southern Normal School was started in 1911 and had an enrollment of 280 students at one time. The school sits on a large campus on the north side of Brewton and has many fine buildings. Kittell Hall was a girl's dormitory with a very nice student lounge. (Courtesy of Yank Lovelace.)

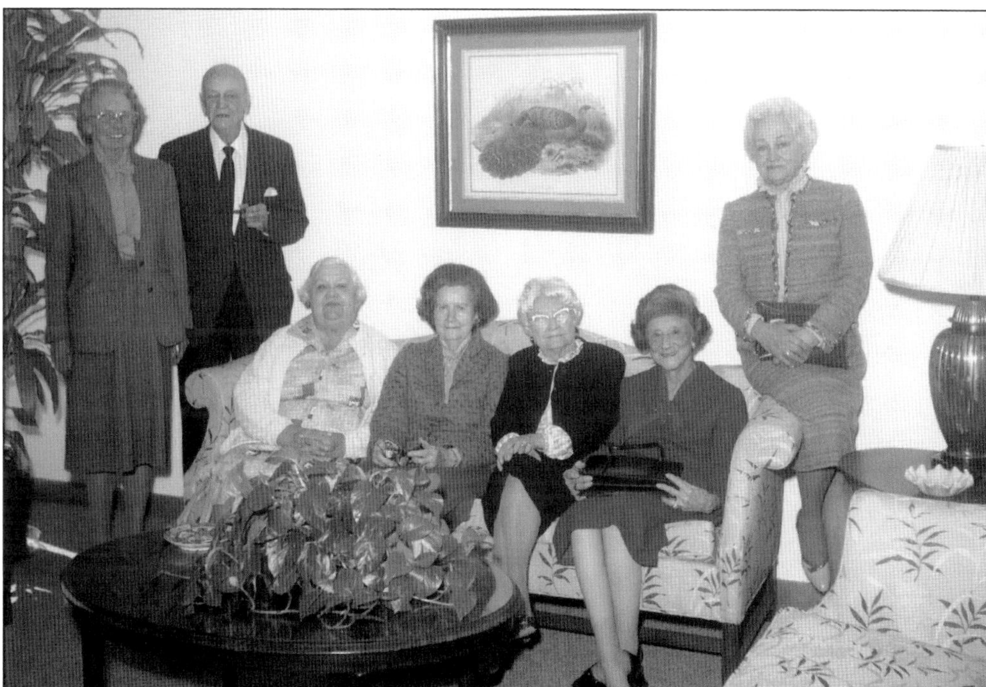

Representatives of the founding members of St. Stephens Episcopal Church appear in 1982. From left to right are Margaret Strain, Al Moss, Jean Moss, Marie Downing, Ethel Sowell, Lucille Miller, and Elvira McMillan. (Courtesy of St. Stephen's Episcopal Church.)

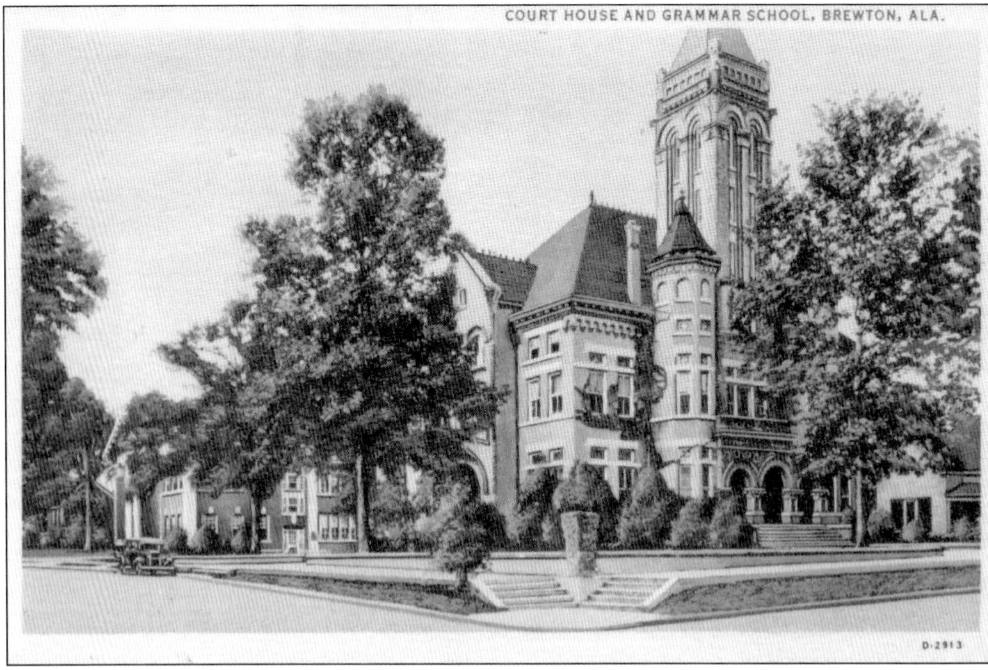

In 1901, the original courthouse in Brewton was condemned, and the movement to build a new courthouse began. The second courthouse was occupied in 1902 and served that purpose until 1960 when the current courthouse was completed. (Courtesy of Yank Lovelace.)

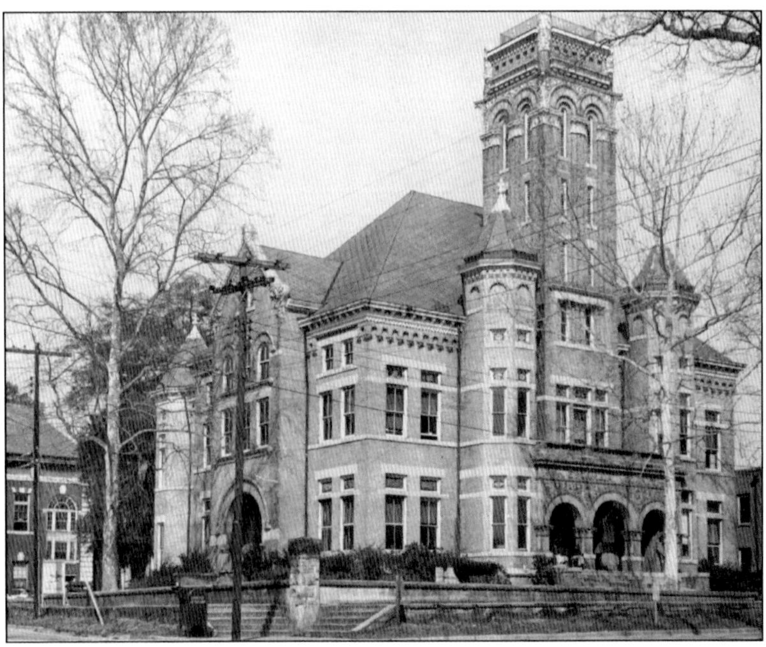

At some point in the life of the second courthouse, the steeple was removed. The Gothic building was in use for 58 years. Only the two cannons on the front lawn that were from Fort Barrancas in Pensacola survived the transition. (Courtesy of Yank Lovelace.)

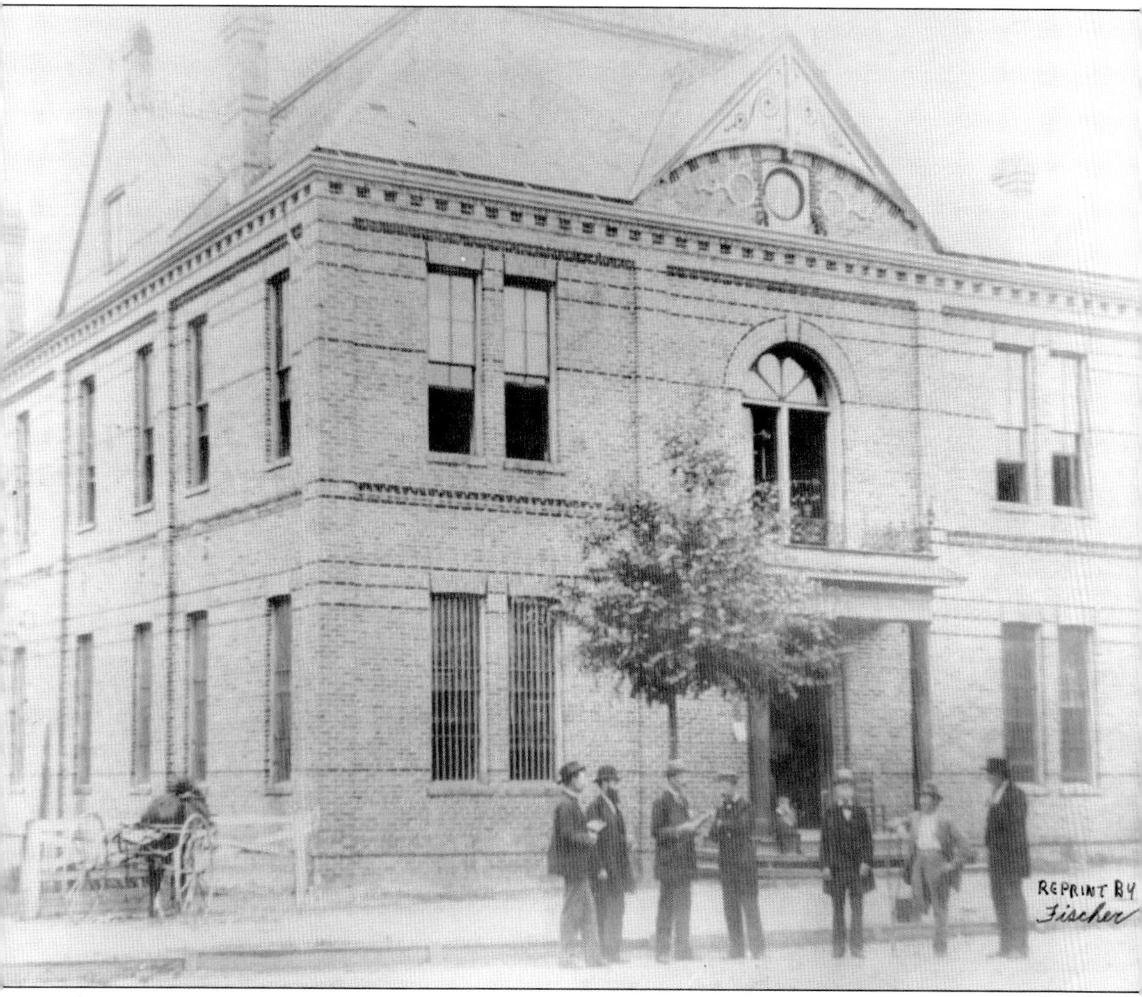

The original courthouse was built in Brewton after the county seat was moved from Pollard to Brewton. This secret move transpired in the middle of the night in 1882. The county officials standing out front are, from left to right, Millard F. Brooks, circuit clerk; Edward S. McMillan, sheriff; Allen Johnson, tax collector; Norville R. Leigh, judge of probate; James Hodges, tax assessor; and James Herrington, treasurer; the child on the steps is Ed Leigh, son of McMillan. (Courtesy of Jerry Fischer.)

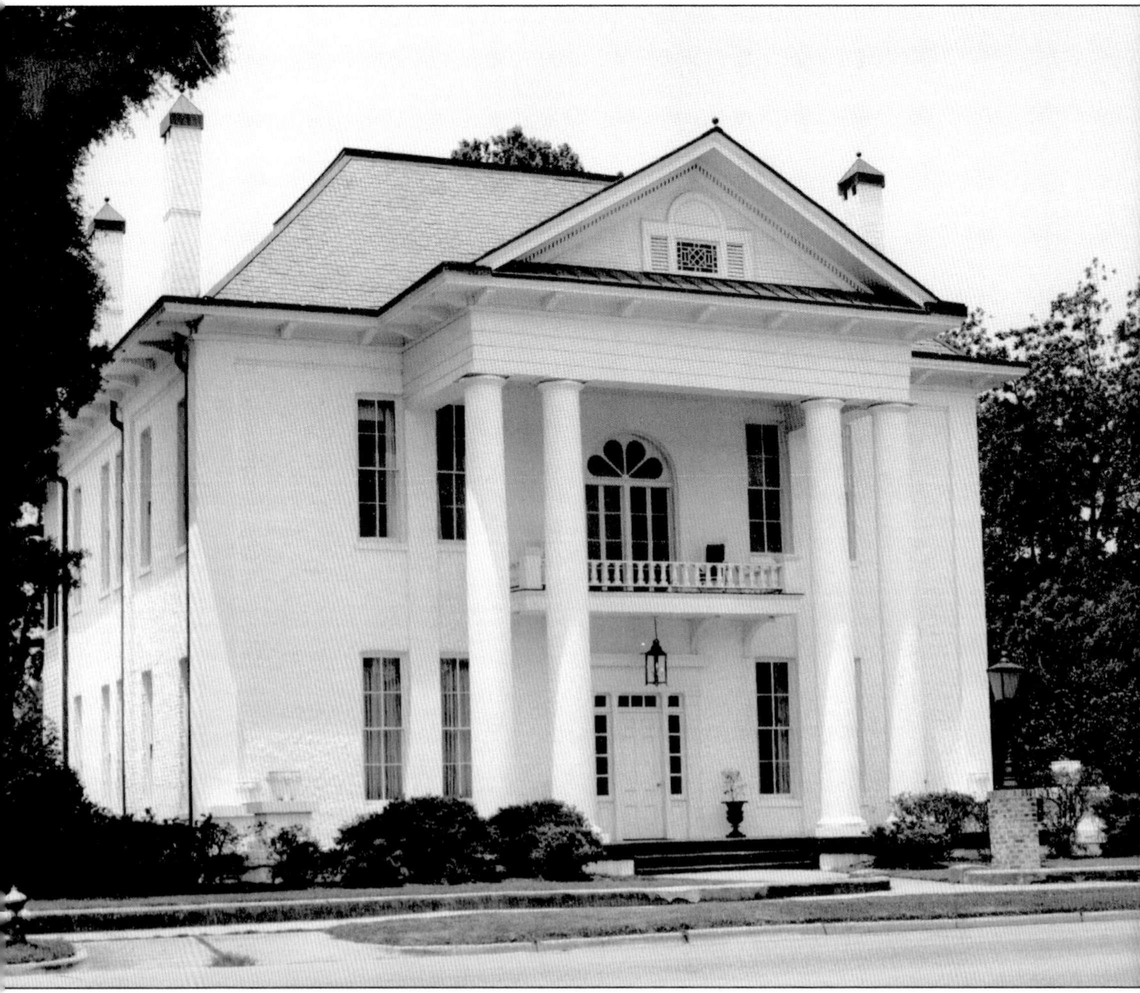

The old courthouse was converted into a residence for a while and then into an office building by Thomas McMillan in 1969. It became known as the Leigh Place and continues to be an office building today. (Author's collection.)

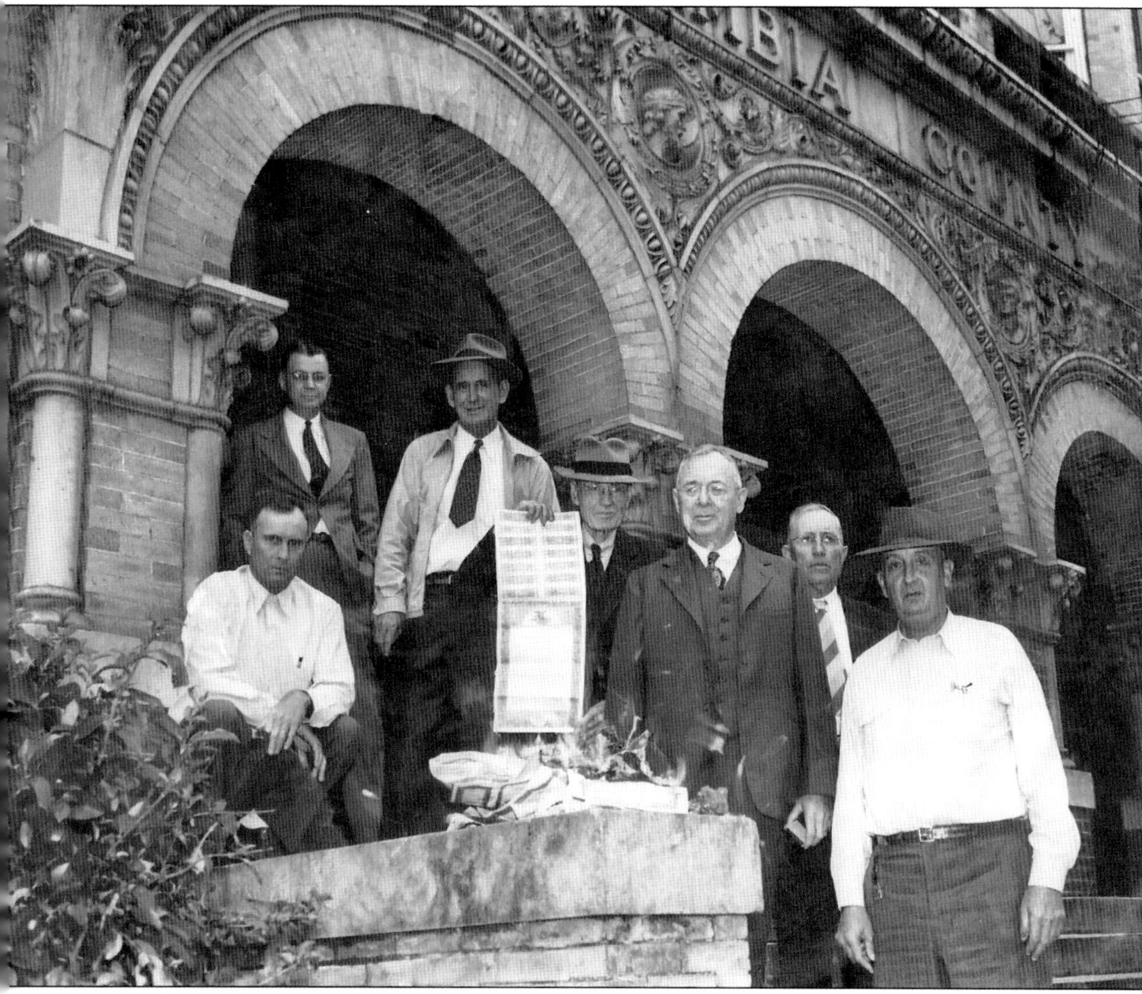

In 1947, the county commissioners held a "burning of the bond" ceremony on the steps of the courthouse. The gentlemen are, from left to right, Dick Fore, commissioner; Wayland Mills, county clerk; J.B. Stallworth, commissioner; unidentified; Thomas A. Peavy, probate judge; M. J. Cook, commissioner; and D.P. Liles, commissioner. (Courtesy of Buddy Liles.)

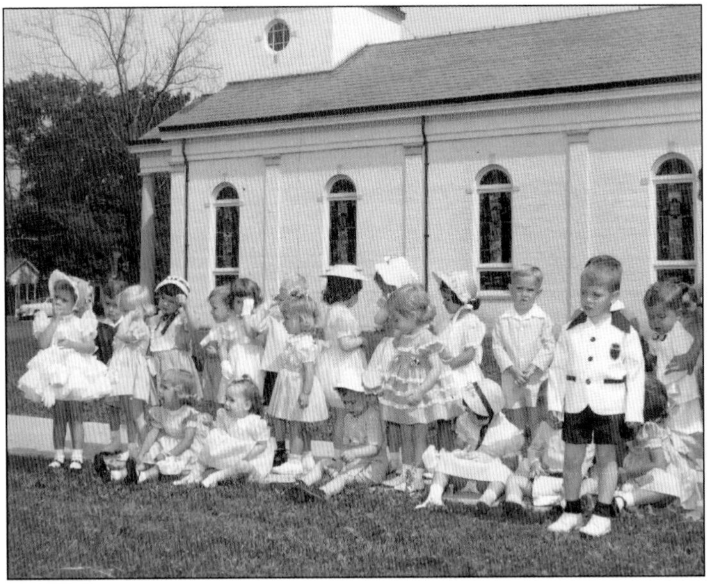

Easter Sunday at the Methodist Church in Brewton in 1951 was a joyous occasion. Everyone was dressed up and ready to go to find Easter eggs—but not ready to look at the camera. (Courtesy of Buddy Liles.)

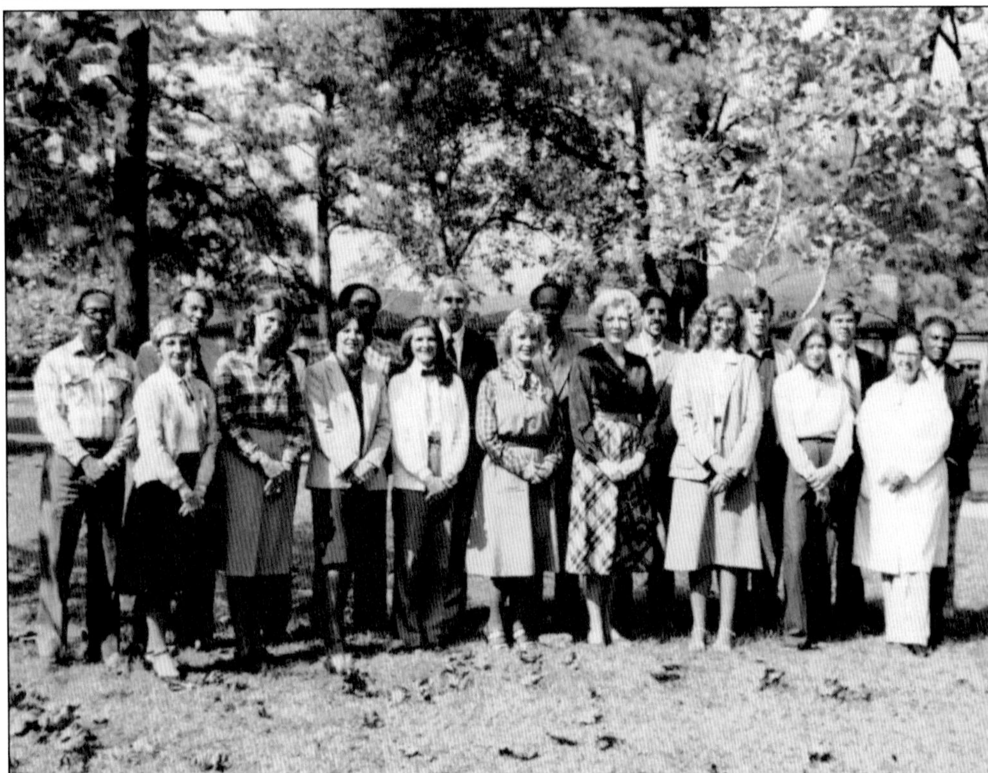

The faculty at Jefferson Davis State Junior College in Brewton, Alabama, stands on the beautiful, wooded campus that is situated in the Alco area. From left to right are (first row) Hermine (Holzborn) Downing, Cindy Moore, Rosemary Jernigan, Glenda Miller, Mary Lee Wells, Ranella (Holley) Merritt, Ann Biggs-Williams, Becky Everage, and Jackie Brock; (second row) Frank Cannon, John Lee, Jack Davis, Eldred Pritchett, Bob Owen, H.K. Matthews Carney, John Cross, Mike Tedder, Mark Reynolds, and Buren Snowden. (Courtesy of D.W. McMillan Trust.)

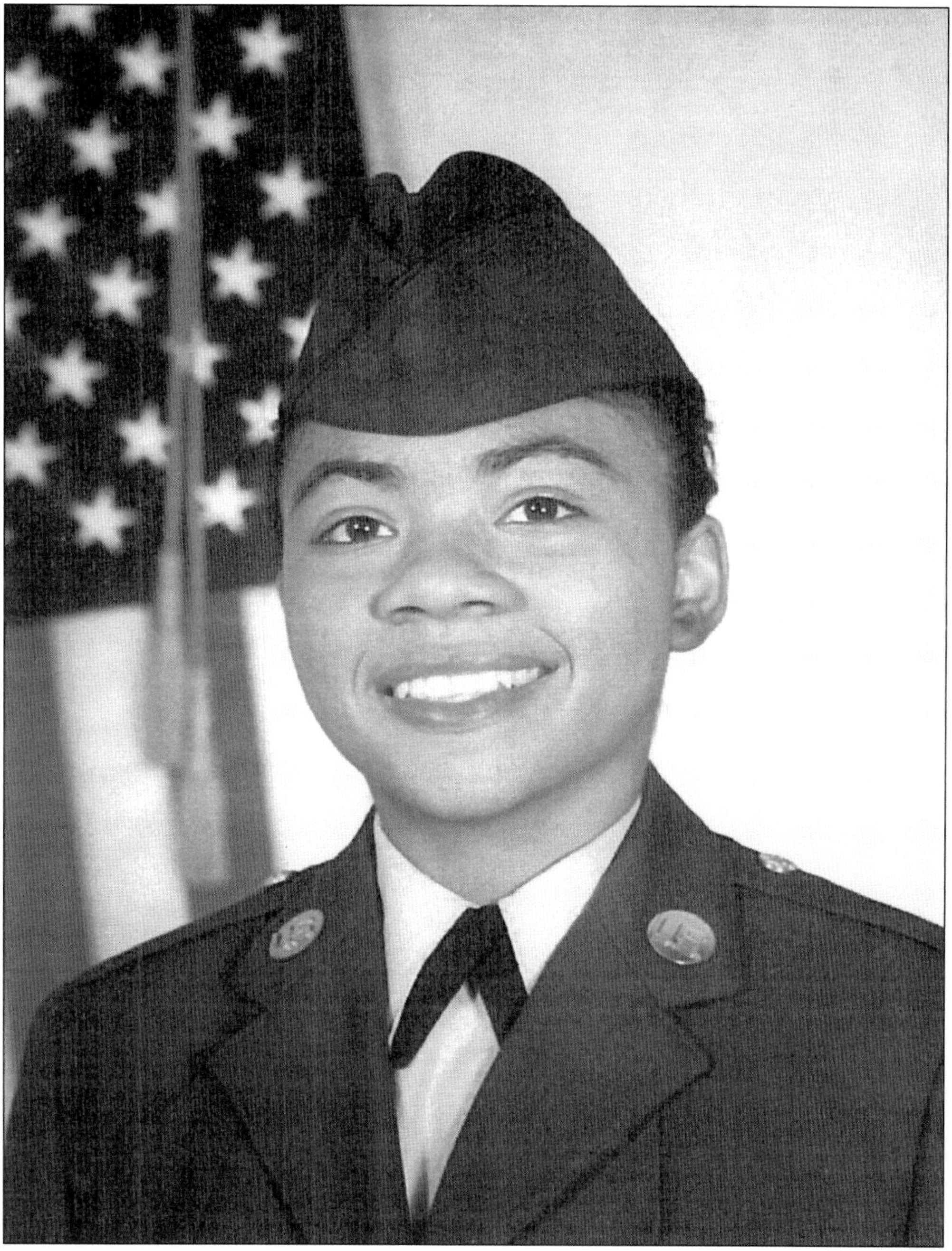

Sgt. Tamara C. Thurman graduated from W.S. Neal High School in 1994. She enlisted in the Army out of high school and served in Bosnia, Korea, and Germany. She was awarded the Army Commendation Medal, the Army Achievement Medal, and the Army Good Conduct Medal. She worked as an assistant in the office of the deputy chief of staff for personnel at the Pentagon. It was at the Pentagon on September 11, 2011, that she lost her life. She proudly served the United States for seven years. (Courtesy of Saundra Woolen.)